SpringerBriefs in Environmental Science

For further volumes:
http://www.springer.com/series/8868

Daniele Fabrizio Bignami

Towards a Territorial Multi-Disaster Buildings' Resistance Certification

Springer

Daniele Fabrizio Bignami
Fondazione Politecnico di Milano
Milan
Italy

ISSN 2191-5547 ISSN 2191-5555 (electronic)
ISBN 978-88-470-5222-2 ISBN 978-88-470-5223-9 (eBook)
DOI 10.1007/978-88-470-5223-9
Springer Milan Heidelberg New York Dordrecht London

Library of Congress Control Number: 2014935973

Printed on acid-free paper

Springer is part of Springer Science+Business Media (www.springer.com)

Acknowledgments

The realisation of this book would not have been possible without the trust and encouragement I was given by Prof. Maria Cristina Treu and Prof. Renzo Rosso and, of course, the support received by the *Fondazione Politecnico di Milano*.

My mother and father, as they know, have been essential. The fresh open-mindedness of my students at the Politecnico di Milano is a motivating charge. Edvige a benchmark.

Allegra and Massimo, with their amazing smiles, have provided a fundamental support at any moment of this research.

Contents

Introduction

> Managing risks from natural disasters requires better management of exposure to natural hazards, through urban and land-use planning. A disaster management policy needs to encompass prevention, resilience and reduction of individual vulnerability and strengthening eco-systems. In hazard-prone areas, property owners will have to invest even more in property risk reduction measures.
>
> European Commission[1]

Despite the huge availability of publications and initiatives concerning knowledge, and subsequently potential actions to be taken, on disaster risk and its consequences, large possibilities of improvement are still in front of us to find ways of analysis and valid methods to implement real large-scale strategies in responding to disasters.

The present book outlines a possible approach for developing an operational new scheme that addresses, specifically, recognised deficiencies and aims at implementing proper actions to obtain an effective risk reduction, especially in the context of built urban areas.

In order to reach this goal, the book offers and illustrates a *multidisciplinary* proposal by uniting and merging different disciplines and fields of study, such as the disaster risk analysis (DRA) techniques, the sustainable land-use and urban planning policies, the environmental assessment methods, the different buildings safety topics, the certification and auditing management systems tools.

Along these lines our text, not thought as a manual, brings together the initial results of a research scheme aimed at testing a new feasible strategy in promoting buildings' safety. The proposal is to achieve a *Certification of the Predisposition of Resistance of Edifices to Disasters—CE.P.R.E.D.* (referred to, in the text, also as *P.R.E.D.* certification or rating/management system). To put it more simply, a more communicative name for the certification was *Safe House* (in Italian: *Casa Sicura*), the title chosen for it when, in 2008, active consideration on this theme began its route thanks to the official and economic support of the Province of Milan.

The idea is still to try out a new tool, an original scheme to be applied in the field of anti-disaster safety measures. Our scheme is meant to offer support to the

[1] Green Paper on the insurance of natural and man-made disasters, Strasbourg, 16.4.2013.

community in defending it from catastrophes. Its basic starting point is the assessment of the multi-risk situation of existing buildings (both from the point of view of hazard and from that of vulnerability). This estimate is to be carried out at different territorial levels and scales: district, province, local/proximity and the buildings themselves. The estimate itself, favouring the building's benchmarking, would thus encourage and guide preventive and protective renewal in built-up areas. Such areas have, with alarming frequency, been struck by catastrophes such as floods, landslides, earthquakes, forest fires, industrial accidents, gas explosions, structural collapses and other natural and man-induced calamities.

Our choice of presenting this proposal of new buildings' certification and rating system in a book is due to the decision of conducting its design and run in, not only by a phenomenological or mathematical/logical point of view, but also through a comprehensive approach including hypotheses and solutions concerning many different aspects of the problems. The aspects under investigation include: the perception and the communication issues related to the needs of a modern disaster risk management (DRM); the audit and the certification management systems' requirements and guidelines presently in force at an international level; the real availability and the exploitability of data bases and maps on different kinds of disasters and related components; the proper ways to carry out the necessary surveys' activities on the ground and "around and/or in" the buildings; the right balance between the effort to obtain final results and the depth of the required examinations; the link with the economical aspect of damages and losses caused by disasters.

The methodological key elements of our work are: a hybrid approach to face a complex field of applications of the risk analysis methods; the introduction of both the logic of the market forces and that of the bottom-up logic on the topics of building safety and disaster risk reduction; a modern communication and perception disaster risk solution; and, in general, a potential and additional new strategy in order to try to contribute to activate effective risk reduction and thus obtaining more sustainable levels of public finances.

As already mentioned, the work was and is still developed in Italy. Therefore what we present and discuss here is a revised, upgraded and critically reviewed work of both theoretical and on-the-field research path, that have already been followed for some years in Italy. The aim of this work is to extend and to validate the results of the investigations at an international level and to try to give an original contribution to the debate regarding the European Commission "Green paper on the insurance of natural and man-made disasters" of April 2013. It also aims at giving an additional point of view, or even a further element, to complete "a strong European Union Disaster Risk Management policies that aim to build resilience against disasters and mitigate their most severe effects" [EC 2013]. As a matter of fact, the PRED certification, here described, could represent a useful answer to complete the EC framework, since it attempts to point out to the omitted or underrepresented aspects of the Green Paper closing question: "This paper addresses specific aspects related to the prevention and insurance of natural and man-made disasters. Have any important issues been omitted or underrepresented? If so, which?".

Along the Research Path

The reflections, in this text, also bring together some significant material from previous studies, which we have adopted and broadened. We refer to:

- The EU project "interreg Medocc" *Qualité dans le Territoire—QUATER* which was finished in 2004. The project[2] (carried out by the Lombardy Region, in cooperation with various departments of the Milan Polytechnic with, as partners, the Piedmont Region, the Parco delle Madonie's Organisation, Spain's Valencia and Baleari Islands communities and France's P.A.C.A. region) was intended to gather material for a collective manual evaluating territory risks as well as showing and controlling the relationships between the various risk-analysis indicators with regard to the critical level of different territory risks—such as chemical industrial areas, forest fires, hydro-geological areas, transport of hazardous goods, coastal erosion and unloading contaminating waste in coastal seas.
- The 5th Civil Protection book *Prevention of risk of collapse in residential buildings*, published in 2001 by the Lombardy Region with the Milan Polytechnic. This book[3] shows an effort to tackle a delicate and complex subject, the risk to buildings, which depends on the ageing of supporting structures and the problems of intervening for safety. This text tried to tackle the subject in simple terms but with scientific accuracy, offering a guide for an individual to make his own definition of the possible signs or "symptoms" of eventual structural faults.

As a further step in this research path, the first pilot project for applying the "Safe House" scheme ran from spring 2008 to spring 2009. It was carried out in Milan Province territory and was possible thanks to the co-financing of the *Direzione di Progetto Diritti, Tutele e Cittadinanze Sociali*, of the *Assessorato Tutela dei Consumatori—Protezione Civile* and also thanks to the working support of the *Assessorato al Piano Casa Metropolitano* of the province administration.[4]

Research and development for the initiative have gone ahead on the lines of the ampler Italian-Swiss EU interreg project[5] *MIARIA—Monitoraggio Idrogeologico Adattativo a Supporto del Piano di Rischio Integrato Alpino* (hydro-geological monitoring in support of the Alpine Integrated Risk plan), begun in 2009 and carried out by the Lombardy Region, in cooperation with, as partners, the Milan Polytechnic (Lecco Campus), the Lecco Province, the Milan Polytechnic Foundation, the University of Italian Switzerland and the University of Applied Sciences and Arts of Southern Switzerland. The plan was intended, among other

---

2 Project web site: https://quater.reteunitaria.piemonte.it.

3 Available at www.protezionecivile.regione.lombardia.it (protezione civile—pubblicazioni—quaderni di protezione civile).

4 Project web site: www.fondazionepolitecnico.it (eventi trascorsi/past events—2009).

5 In the framework of European Regional Development Fund—ERDF. MIARIA project web site: www.miaria.regione.lombardia.it.

aims, to "coordinate and improve the efficiency of the whole system of risk—and emergency—management". Second, it aimed to "strengthen the public's involvement as to sensitivity to and perception of risk—with regard to self-preservation in emergencies and in land use". These aims are to be achieved through dynamic scenarios of integrated risk in mountain settings so as to plan intervention and control emergencies and, on the whole, strengthen resistance to disasters.

Milan, September 2013

Chapter 1
Territory, Buildings, Sustainability and Disaster Risks: The Necessity of New Answers

1.1 The Roots of the Problem

How safe is the building I am in? A question like this is what every member of the public probably asks himself when he comes across media articles describing the catastrophes that continually strike conurbations on our planet.

But can we define the risk level of our private home or the public buildings where we spend our day? And are there technologies that could help? At the moment, such information is very hard to come by. But if we could get hold of it before buying a house or judging what restructuring to do (or before beginning a school year!) it would be extremely useful.

Here we shall focus on the crucial theme of the necessity of reinforcing the relationship between the land planning and the edifices' disaster risk reduction activities. One of the most relevant outcomes of our work resulted in the need of an integrated approach to the subject matter. As part of the community of researchers in the environmental sciences, and wanting to put theory into action, the most difficult problem was (and still is) to identify, and subsequently offer operational tools useful for the implementation of our integrated approach. In other words, the pivotal point today is still to propose functional solutions to the above questions according to this integrated view; our proposal could be one of the modern ways to trigger a modern mechanism capable of overcoming the situation of waiting, or inaction, in our fight against disasters' effects and consequences.

1.1.1 Overcoming Limits by Taking Action: Simulating Stress Tests Also for Buildings

In need to practice a competitive sport, an athlete has to undergo specialist medical inspections at regular intervals as is determined by his sport's medical body if he is to stay qualified. This is in order to minimise health risks or harm during competition or training. Such inspections are different from those usually given to

D. F. Bignami, *Towards a Territorial Multi-Disaster Buildings' Resistance Certification*, SpringerBriefs in Environmental Science, DOI: 10.1007/978-88-470-5223-9_1,

amateurs to certify their health and fitness. A professional inspection includes some tests carried out "under stress" so as to check the organism's response to fatigue. These tests are not the same as our daily movements of physical effort.

In the past, ad hoc laws and regulations have fixed hygiene, health and safety standards for buildings (above all in the richest countries). Such standards have intended to reduce possible accidents as regards both basic functions and further risks such as fire. These norms have for some time, in a generally satisfactory way, ensured daily activities for hygiene and the commonest bad weather conditions. But very often as the years have passed and many new factors have come into play, these norms or standards, generally, no longer ensure a regularly updated contextual evaluation of the strength of a building when subject to a calamity and therefore to extraordinary stress.

We can list the factors that compromise the protective performance that buildings should offer. This list is long and, among other facts, includes the original inadequacy of the norms and the construction's features, obsolete design decisions, ageing structures, norms never updated, structural imperfections, degradation of buildings due to misuse, risky modifications, uses unforeseen by the original architect, as well as unpredictable development in the status of buildings and localities. Whatever the situation, there are many "new" or "regulated" buildings that should have been (but were not) able to stand up to all the quakes that have happened, for instance, in Italy from the Seventies till now, when the first real anti-quake norms came into force. Similarly, today's regulations do not give enough enforcement to the checks, the care, the maintenance and (if necessary) strengthening of *existing buildings*—even though these often have features that are anything but satisfactory. And in spite of this fact, these buildings form the vast majority of Italy's inherited constructions. Probably, the situation may be quite similar in many disaster-affected portion of our planet. New regulations concentrate, above all, on new structures.

Exaggerated trust in modern building techniques, the fragmentation of specialist knowledge, of competence and responsibility, the way too many project aims depend on minor importance market choices when faced with scarce or hard-to-get resources, all those factors taken together with the supposed variety of calamities and crude overestimate intervention costs have led to an outstanding effect. The first consequence was the delegation of responsibility, and then to lose, or turn a blind eye to, what remains one of our basic instincts—testing and maintaining the safety of our own habitation, our "shelter". Such an instinct, in a modern version, has to be recovered and strengthened.

The main aim of this research is to achieve the above. Our paper illustrates the work that was involved in inventing a proposal for eventual certification of buildings' strength against disaster risk. The certification would be an integrative element of the current regulations and codes (not necessarily legislated) thanks to an *innovative methodology*, which would be non-invasive, based on precise collection, analysis and organisation of existing information and knowledge.

1.1.2 Our Perspective in the Framework of Initiatives to Reduce Disaster Risks in Support of Civil Protection Activities

The reflection and research illustrated in these pages lead to the proposal of a new voluntary system of evaluating risks to buildings and urban assets. Thus, these studies chiefly initiate from the combination of three research fields.

The first and primary of these regards strategies and methods for *reducing the disasters risk* (or Disaster risk management—DRM) which in Italy is dealt with by the National Civil Protection Service (and, similarly in EU, for instance, by the coordination of the Emergency Response Centre—ERC of the Humanitarian Aid and Civil Protection DG of the European Commission; in the USA, by the Federal Emergency Management Agency—FEMA). The second relates to the long and mid-term *land-use planning* policy for built-up areas, to be reinforced as a fundamental tool to the territorial safety. The third concerns *assessment and certification techniques*, because of a few critical aspects of its implementation to the edifices and their territories. The latter proposal aims to define, as far as it is verifiable, the potential disaster-resistance of the building investigated; this potential resistance would be certified as predisposition in defence against eventual calamity phenomenon of a structure during a disaster.

In short, the idea is to invent a project scheme that starts from certification methods for the *product* (cf. Vertinsky et al. 2000; Foglio 2008) and for minimising disaster risk from an up-to-date multi-risk viewpoint (cf. Kron 2005; Grünthal et al. 2006; Ritchey 2006; Schmidt et al. 2011). The purpose of these methodologies is, through well-defined *indices*, to simplify and summarise the performance, in a specific functional situation, of a particular good so as to ascertain its conformity to a precise discipline. There are significant examples of such methods applied to the building industry: e.g. the 'energy' certifications of buildings (cf. Dall'O et al. 2008; Pérez-Lombard 2009) such as, in Italy, those emitted by Bolzano Agency *CasaClima* (www.agenziacasaclima.it) or by the Lombardy-Region *CENED—Certificazione Energetica degli Edifici* (www.cened.it) scheme, or the environmental certifications of buildings such as the US *LEED—Leadership in Energy and Environmental Design* (www.usgbc.org/leed) or the UK *BREEAM—Building Research Establishment Environmental Assessment Method* (www.breeam.org) systems.[1] Nevertheless, as we will show in the following pages dealing with Disaster Risk Management, we are facing a more

---

[1] Other relevant examples of energetic or environmental sustainability systems are (inter alia): the Swiss *MINERGIE* sustainability standard; the German *Passivhaus* energy standard; the German *DGNB—Deutsche Gesellschaft für Nachhaltiges Bauen* sustainability system; the French *HQE—Haute Qualité Environmentale* green building standard; the US Energy Star (energy); the Japanese CASBEE (sustainability); the Canadian Green Globe (sustainability); the Australian Green Star (sustainability).

complex multi-dimensional problem requiring knowledge and experience from a wide range of disciplines.

The proposal of our certification has been developed at the Milan Polytechnic on the basis of collaboration between the Milan Polytechnic Foundation, the Architecture and Planning Department and the Department of Building Environment Science and Technology. This certification is meant to safeguard the public, the "consumers" of the built "good", by means of an evaluation, rating and labelling system. The proposal tackles the problem of spreading knowledge and awareness of buildings' safety. This is to be done in various ways achieving in the end a *labelling* system with "*Blue shields*"—from one to five shields indicating the construction's resistance strength against disasters.

Damage brought by a disaster does not only depend on the phenomenon's violence but also on the building's structure features and its locality. The proposed certification takes into account a variety of risks and events. These include earthquakes, landslides, floods, avalanches, weather extremes, industrial and hazardous-goods-transport accidents, forest-fires, etc., as well as the technical nature of the building material used, of the inbuilt installations and their maintenance and conservation. The gathering and analysis of all this information available to us mean we can more precisely single out the high-risk land areas and most vulnerable structures—and as a result we are able to give the public and the local administrators more information and awareness.

The chosen target for observation and inclusive evaluation of these interactions is the building itself—since it is here that the considered events concentrate. Such events may have international import like hurricanes or floods but they may also derive from urban or suburban factors such as industrial accidents or forest-fires, or from individual sites or structures like landslides or gas-leak explosions. Secondly, the building (house, home, refuge) is what attracts the individual's greatest attention in its function as a basic constituent of human presence where the owner or administrator is responsible for his own safety and that of his family or other dwellers there.

Thanks to this undoubted role, the building as structure can itself become the initial and guiding impulse to achieve, now at last with a telling impact, what has been called, in Italy, the "*greatest of all public works*" (Gruppo 1996, p. 183), meaning, with a *logical strategy based on priority maintenance and prevention*, the creation of all-round safety for the country or, rather, a *general increase in the nation's safety*. Similarly, the PRED certification's approach may represent, within the scientific on-going debate at the European level, an important component in which it is underlined the need that persons endowed with policy decisional powers may be able to prepare the ground for such policy changes, well ahead of the calamitous events which normally would only be able to put them in force with immediate effect (Penning-Rowsell et al. 2006); it can also be seen as one of the alternative proposals to the environment and economy integration, either for climate policy improvement or for strategic environmental assessment activities (cf. Murphy et al. 2000; Kuik et al. 2008; Therivel 2012).

The approach taken in the clearly fundamental and indispensable territorial risk-mapping, also an essential preparation for the scheme described in this text, was not on its own successful, i.e. not successful in creating the conditions essential for converting the acquired knowledge in active intervention. *CEPRED* thus could constitute one of the basic elements for finalising and activating the information for acquired as well as energizing action in tackling a problem that is, by now, well-known but to solve which, action has been slow-paced both collectively and individually—and not only in Italy.

As well as detailing our proposal and research, a complementary but not secondary aim of this text is to help to revive the debate on feasible interventions to reinforce the *safety culture* of buildings and thus of the whole country (and its impact as overall disaster risk reduction culture). This debate should take place among institutions, owners, building-sector and insurance representatives and the bodies involved in protecting the land from disasters. Nowadays, the debate seems too often concentrated on the problem of reacting to emergencies or on the difficulties of restoring the localities affected by disasters. Instead, we should be dedicating the same amount of attention to a well-timed *reducing* and *not increasing risk.* This would encourage territory and town-planning as well as a more knowledgeable and innovative building industry. It would also lessen the damage and the mourning that future calamities will inevitably bring. In this way, our work also proposes itself as a contribution to one of the most relevant debates on the topic of disaster risk reduction, highlighted by the European Commission "Green paper on the insurance of natural and man-made disasters" of April 2013. The Safe House (Casa Sicura—PRED Certification or Rating System) initiative can, in fact, have a direct effect in creating attention on structure and land maintenance, which should be formative and stimulating in rendering wider reaches of society and institutions more conscious and responsible—without going against market mechanisms but, in the long run, helping to save public and private wealth.

1.1.3 Do We Need New Strategies to Protect Our Houses from Catastrophe?

In Italy, just a few days after the 2009 earthquake of L'Aquila and Abruzzo, an increasing debate was started on the possibility of knowing in advance the resistance of a building, before being ravaged by a calamity. Our work, in its first "version" had just been presented a week before.[2] Naturally, it was not yet ready to be applied on a large scale but that debate encouraged us to continue our

[2] The meeting to present the first project results was held on 31st March 2009 in the Audition of the Milan Polytechnic, via Pascoli 51.

research. However, here we want to show the framework on the basis of which our certification idea can be logically pursued.

All over the world, just as in Italy, one observes, time after time, calamities and disasters, small or great, caused by nature or human action. Such events now continually punish large numbers of inhabitants. The seriousness and frequency of these events (for various reasons linked to the excessive consumption of resources) has grown substantially (cf. EC 2013). According to *business-as-usual* scenarios, the increase rather than decrease in such developments (cf. WB and UN 2010; Munich Re 2011) is in contrast with the common idea of *progress* understood as bringing benefits from every viewpoint. Owing to the concentration of human activities, the above situation has reached such levels that, more often than not, it compromises the property (riches and income) of the calamity-struck places for a long time (Guha-Sapir et al.—CRED 2011). Our reader may be a supporter of the *pessimistic* or *strong-sustainability* model or of the opposing *optimistic* or *weak-sustainability* one (cf. Simon 1981; Lomberg 2001; Neumayer 2003) as regards the evaluation of human influence on both the ecosystems and the human systems of our planet. In either case, our reader will share the urgency, or at least the desire to speed up the introduction of practical feasible counteractive mechanisms able to connect the above tendencies—at least in advanced-economy countries. The ideas presented in this book, furthering a *certification, rating and labelling system* of the strength of buildings under threat of disaster, spring from the specific aim of creating a counter-effect mechanism apt to trigger a positive feedback loop.

As one of the most important theoretical basis of the research carried out, we have (nowadays typically for the research field we are dealing with) Tietenberg's and Lewis's clear statement in 2009 that "*market forces are very powerful. Attempts to solve environmental problems that ignore these forces run high risk of failure*". These forces undoubtedly also affect the building market and its construction areas, i.e. building plots. This is why this paper of ours has not taken the traditional *authoritative* approach (i.e. with rigid regularity) to the problem of buildings' safety and decision-making. Instead, it points towards a more up-to-date collaborative and proactive approach (IPCC 2001; Penning-Rowsell et al. 2006; O'Brien et al. 2006; Geneva Association 2009). Our approach, then, aims at producing a flexible and innovative procedure thanks to the voluntary choices of individuals and citizen groups, without having to wait a long time for public or bureaucratic discussion and decision, which often risks rigidity and failure. Our hope is to tackle the problem of buildings' safety with new tools. Because of its dimensions, this safety problem has not been adequately dealt with by the *top-down* interventions attempted by governments in Italy and elsewhere. Such inadequacy has been present even in rich, advanced countries like Italy because of the well-known and complex problem of the scarce resources applied to making controls efficient. This failure regards inefficiency in improving and expanding public sensibility to the problem, with the result that we have limited and isolated initiatives, compared to the needs, of the prevention, protection and mitigation from disaster risks.

1.2 Built-Up Land Area Safety: A Fundamental Component of Sustainability

Faced with the continual calamities that have struck our planet, public opinion reacts with profound worry and anxiety. The disastrous effects on our homes and workplaces, our hearts and minds, mean that the problem of protection and safety in areas of human dwelling can no longer be put off. Planning initiatives to safety have become urgent—unless we are ready to accept the foreseeable consequences with the likelihood that in the long run they will certainly worsen.

Like many other countries, Italy is reasonably rich and has a high education level so, nowadays, it possesses the competence and ability to deal with the problem of improving its buildings' resistance in case of calamity. The choice and decision to do this means beginning to tackle the outstanding question regarding policies and strategies meant to limit personal tragedies and material damage brought about by recurrent disasters.

This question is also complicated by its ethical and social-cultural content and its notable technical and economic effects. Success in sensitizing and guiding town-planning, land-protection and the building industry as well as restructuring to decrease risk in new and old buildings produce various long-lasting effects (cf. Lekuthai et al. 2001; Guzzetti et al. 2005; Messner et al. 2007; Jonkman et al. 2008; Thieken et al. 2008; Barredo 2009; Merz et al. 2010), such as:

- reducing deaths and physical or emotional suffering and the material and/or "tangible/intangible" damage, all directly caused by disasters;
- reducing damage to artistic heritage in architecture, landscape, galleries, museums, archives—all treasures that cannot be reproduced. This is particularly true for Italy as well as for other European and Asian countries;
- greater protection of national inheritance in the shape of buildings and urban layouts which represent years of work and saving by many generations as well as the preservation of the social functions of such sites. This would be defence from the "indirect" damage due to loss of activities and lack of weather protection.

As regards, at least, the first two of these points, one may suppose there is quite widespread awareness—even though not yet strong enough to trigger quicker policy improvement. In these pages we also hope to show how the catastrophes are closely, more than it is usually supposed, connected to economic and social activity—a connection whose real implications are too often neglected in favour of the mass-media spectacular reports which are useless in inducing real action and strategy and whose, for example, estimate of the damage done prove hasty and excessive emitted as they are too soon after the event. By our choice of information and strategy with the PRED certification, we hope to define these aspects more clearly and strengthen the premises for reasonable and effective action.

The increase in disaster phenomena has been given more and more widespread attention (as proved by the UNISDR initiative *Hyogo Framework for Action—HFA,*

2005–2015—Building the resilience of nations and communities to disasters[3]). This increase is due to different mechanisms depending on the continuous growth of the human population on the planet and on the conspicuous consumption of our various natural resources. And this despite the fact that developing construction techniques have led to notable advances in buildings' performance achieving high levels of *comfort* from different angles (thermal, acoustic, lighting, hygiene, hygrometric, etc.). As a matter of fact, these successes now find themselves challenged by demands that are apparently new but were instead requirements originally under-valued in the developing days of building technique and practice when needs were envisaged as short-term rather than medium-term or long-term. The foresight involved should have been extended to the whole life-cycle of the buildings, often bearing in mind the *periodicity and recurrence* of grave or medium-grave disasters.

We can point out, more clearly, a series of phenomena that, with at times dreadful results, have intensified natural territorial weakness such as:

- extending urbanisation, seen as growth in catastrophe-exposed areas, without in advance doing adequate investigation and planning of the locality and the study of the intervention features needed;
- as a consequence, we have an often irrational increase in the building occupation of soil and natural free land areas;
- the increase and scattering of population going together with the growing demographic concentration *in loco* of new settlements;
- the rise in wealth and property values increase the exposure and vulnerability of economic assets and, consequently, the recurrent severity of losses;
- such developments lead to the increasing impoverishment, degradation and over-exploitation of natural resources (water, forest, extraction of raw material, etc.);
- global climate change process (warming and its effects, including water cycles) (IPCC 2012).[4]

In developing countries, where the above developments are joined with poverty, low education levels and feeble infrastructure, the consequences can be fatal not only in the present but also for the future overall development of entire countries (Unisdr 2008). As an example valid for all, take the recent Haiti earthquake (January 2010), where the disaster cost was 8 billion dollars, i.e. more than the country's total gross domestic product (GDP)—the final cost was equal to 128 % GDP (Guha-Sapir et al.—CRED 2011).

---

[3] http://www.unisdr.org/we/coordinate/hfa.

[4] Intergovernmental Panel on Climate Change (2012), *Changes in Climate Extremes and their Impacts on the Natural Physical Environment* in *Managing the Risks of Extreme Events and Disasters to Advance Climate Change Adaptation*; European Environment Agency (2010), *Mapping the impacts of natural hazards and technological accidents in Europe, an overview of the last decade*.

Similarly, according to leading insurance companies, catastrophes are getting more frequent and consequent losses are increasingly serious also in advanced countries. Only in 2012 Munich Re NatCatSERVICE signalled that overall losses throughout the world due to natural disasters reached the amount of 170 billion of US dollars (Munich Re 2013). That is true even if measuring and recording data on disasters is a very difficult work, often underestimated, since good quality reporting of economic losses is still limited to 30 % of calamity events (Smith 2013). In fact, though generally the vulnerability of buildings decreases as the country and its wealth grow, this process is not usually quick enough to make up for or even reduce the exposures to disaster risk. The number of deaths in advanced countries is less shocking, but its impact in context is no less significant. The L'Aquila's quake of April 2009 is a recent and emblematic case (cf. Erbani 2010; Bignami 2012).

The slowness of human reaction in facing this problem seems unaffected by the usual means of intervention applied by the various authorities involved—even in super-rich countries with an efficient civil service. Floods and tempests find us unprepared in both north and south Europe (not to speak of east and west), e.g. *Winter Storm Joachim* striking France, Germany and Switzerland (15–17th December 2011), and *Xynthia* in Portugal, Spain, France, Belgium, Luxembourg and Germany from 27th February to 1st March 2010 as well as the 2002 and 2010 floods in Germany, Austria, Czech Republic, Slovakia, Poland, Hungary, Romania, Ukraine, Serbia and Croatia. In the same way floods, hurricanes, severe storms and tornadoes struck repetitively in 2012 also in North-America (*Sandy*, for instance, from 24th October to 31st) and China (floods from 10th May to 24th and from 21th July to 24th). In 2011 floods struck two times in Thailand (in August and in December) and in Brazil (in January). Moreover, the seismic vulnerability of a nation's buildings built without respecting modern rules, regulations and techniques is a problem both for those countries with impetuous growth, like Turkey, and for those with more or less mature economies like Italy, Japan and New Zealand or for those in major crisis like Greece.

Thus the total of deaths, damage and destruction caused by natural disasters is quite often able to spoil the growth prospects of entire countries or regions or even seriously to undermine their accumulated wealth and possessions. This applies especially to the destruction of a nation's building patrimony, buildings which represent a very significant part of a country's or territory's wealth. Bearing in mind the stock of dwellings and structures in the most developed countries, it seems logical to stress how urgent it is to intervene drastically if we are to improve buildings' resistance to disaster.

Such improvements are nowadays more and more often called for by a country's inhabitants—though in many cases only when disasters have just happened or are imminent. The long sustainability of a community and its continuing development can be seriously affected by wrong choice of locality, mistakes in construction techniques and materials, lack in maintenance, failures in checking and political or professional dishonesty. Every community under threat of recurrent disasters must learn to deal with the resulting impacts not only from an ethical, human point of view but also from an economic and ecological one (ecological, this time, in its purest sense[5]).

Promoting the *sustainability* of a territory as regards its environment and its living standard and wealth must include its ability to deal with disaster risk and hold them in mind in reaction programmes—otherwise there is a lamentable and striking gap in the country's preparation. To construct a nationwide mass of buildings which is left open to serious damage in a near-future catastrophe (even though the latter is a random event) signifies a *violation of the sustainability* that should, for future generations, at least be maintained at its present level. The imprudence shown up to now by some generations will be, unfortunately and inevitably, having harm to future following ones, with deaths and the destruction of wealth and history. Sustainability violations have, for some decades, been made in some countries when technical and economic progress has brought a general increase in well-being. Such well-being could have (and might still) bring with it foresighted policies—all the more so since in our times we no longer take for granted the idea that new generations will create more wealth than previous ones.

In that regard, studies by the United Nations show how GDP may not be able to recover its pre-disaster levels. In various cases, this happening shows itself in a persistent drop below the regular pre-disaster level (UN and WB 2010).

1.2.1 The Underestimation of Our National Debts: A Debt of Territorial Safety as a Violation of Intergenerational Solidarity

The situation, as it stands, points with probabilistic near certainty to the fact that in a country like Italy (but not only), national debt will, in the next few years, be increased by the costs of the damage done by new calamities—unless they are taken as straight comparable losses in wealth and assets. The *territorial unsafety* fostered by imprudent construction techniques and careless use of building sites brings into action a special dynamic, i.e. it seems that those now enjoying the nation's wealth and heritage have borrowed future generations' resources that would have covered the losses caused (paradoxically) by the previous generations!

Thus, estimates on the steps needed to pay off the national debt, or trend analyses of the comprehensive wealth of a nation, have to be taken as calculated by defect since they do not take into account the *debt of territory safety* which will have to be paid by generations succeeding those who have caused it.

The above reflections show up even more clearly that there has been a violation in *intergenerational solidarity*—at least in those economically mature west

[5] From ancient Greek οἶκος (house, dwelling, home) and *λόγος* (word, discourse, story or reason, intelligence, opinion, evaluation, thought), so, *thought about inhabiting*.

European countries where there is no longer any guarantee that future generations will be richer than contemporary ones.

Unfortunately many countries cannot any longer persist in covering a large share of the costs of risk reduction and public relief by continuing to compensate victims generously as it was done in the past. Because this is not only likely to aggravate national financial problems but even more likely to "promote" undesired developments in risk-prone areas. Combining modern and well-organized ways of solidarity with building improvements and new standards, partially corrects the imperfect internalisation of risk and increases general efficiencies.

In an another way, countries must control building development in risk-prone areas and must encourage collective individual responsibility for reducing losses and deaths; and they also have to put aside resources against losses timely. It does not matter if this is done by a public mutual system, private market-based system or a hybrid one. As a long-term insurance type contract, with clear risk-based pricing and premium discounts for risk reduction interventions, could strengthen financial incentives by making investment in risk reduction advantageous for both insurer and insured, in the same way it is possible also for State's coverage and tax payers.

1.3 Current Defence Strategies Against Disaster Risks: Brief Overview

Faced with this scenario, nobody can deny that Italy, like many other European countries (as an example of a more or less homogeneous and cooperating community of rich and knowledge-based countries or states), has tried to form bodies for territory planning and managing and for dealing with disasters—and these attempts have been in their aims widely diversified and, on paper, often fairly complete.

First of all, we must cite land use and urban and town-planning systems and those for regional and sub-regional territorial planning (for instance, in Italy regional territory plans and provincial territory coordination plans, including landscaping) as well as those bodies at town-council level involved in direct urban space management (even operational and negotiated urban planning). All these bodies concerned with detailed control of territory growth and dynamics are the very offices that should be the final destination for all the existing ideas on territory safety (Treu 2009); in this way, we consider land-use planning as part of a disaster risk management policy. Furthermore, we will proceed with systems devoted specifically to reaction to disasters; these, too, may be *multi-risk* or *sectorial* (Bignami 2010).

For instance, the first one among the systems relating disaster risk was introduced in Italy some twenty years ago by Law 225 of 1992. These *multi-risk* systems are *communal, provincial and national civil protection (or emergency) plans*, just like *foresight and prevention programmes*.

The second (numerous) group of systems concerns itself with specific hazards, in particular (although the following list is not exhaustive):

- national *seismic zoning* (the competence of the national authorities cf. Solomos et al. 2008),[6] norms for building in seismic zones (on the basis, for instance, of the European Standard Eurocode 8—EN 1998-1: 2004 Design provisions for earthquake resistance of structures) and seismic *micro-zoning*[7] at town level;
- *river basin management plans*, through, for instance, the norms for the risk exposure plans for hydro-geologic risk areas, the *flood protection programmes* and the *hydraulic risk management plans*, the safety norms for *dams* and the *charts for probable avalanche localities* (in Europe combined with the national application systems of the EU Directive 2007/60/EC on the assessment and management of flood risk, having a clear set out timetable of the three stage-implementation process to be concluded by 2015)[8];
- national application systems about appropriate distances from establishments at risk and *external emergency plans*, on the basis of the "Seveso II" EU Council Directive 96/82/EC of 9 December 1996 on the control of major-accident hazards involving dangerous substances, as extended by the Council Directive 2003/105/EC (considering that a "Seveso III" replacement directive 2012/18/EU has entered into force on 13th August 2012 and EU Member States have to transpose and implement it by 1st June 2015)[9];
- United Nations Committee of Experts (COE) on the Transport of dangerous goods basic rules (including those on road and rail), the European ADR and RID transport of dangerous goods regulations (on road and on rail respectively)[10] on the basis of the EU directive 2012/45/EU of 3 December 2012 adapting the annexes to the EU directive 2008/68/EC;
- Wild-fire prevention or suppression and emergency plans (even in the case of wild land-urban interface), land registers for fire-burnt areas.[11]

---

[6] For instance, in Spain: Ministerio De Fomento—19687—Real Decreto 997/2002, 27 de septiembre, por el que se aprueba la norma de construcción sismorresistente: parte general y edificación (NCSR-02); in France: Arrêté du 29 mai 1997 relatif à la classification et aux règles de construction parasismique applicables aux bâtiments de la catégorie dite « à risque normal ».

[7] For instance, in Italy *homogeneous micro-zone charts in seismic perspective*, based on the "Indirizzi e criteri generali per la microzonazione sismica" developed by the Department for civil Protection and the Conference of Regions and Autonomous Provinces (2008); in Switzerland: Principe pour l'établissement et l'utilisation d'études de microzonage en Suisse, Directives de l'OFEG (Office fédéral des eaux et de la géologie), Berne 2004.

[8] For instance, in Italy: Piani stralcio per l'Assetto Idrogeologico (Legge 267/1998 and DPCM 29 settembre 1998); In France: Plans de prévention des risques naturels—inondations, avalanches, mouvements de terrain (loi 565 du 22 juillet 1997 and 101 du 2 février 1995).

[9] For instance, in France: Plan de prévention des risques technologiques (loi 2003-699 du 30 juillet).

[10] ADR—European agreement about international transports of hazardous goods by road; RID—Rules about international railway transport of hazardous goods.

[11] For instance, in Italy: Piano Regionale per la programmazione delle attività di previsione, prevenzione e lotta attiva contro gli incendi boschivi (Legge 353, 21 novembre 2000—Legge quadro in materia di incendi boschivi); in France: Plans de prévention des risques naturels previsibles d'incendies de forêt (loi 565 du 22 juillet 1997 and 101 du 2 février 1995).

These measures are accompanied by many initiatives for *collecting and digitizing information* (including historical connections); such initiatives are very useful in combating disasters. On this theme, in Italy (and similarly in most of the world developed countries), we would like to list the following as significant examples: charts of soil usage, geological and geo-morphological maps, digital models of land-areas, satellite data, (nowadays more and more precise and differentiated), the AVI catalogue by the CNR (National Research Council) on the areas struck in the past by hydro-geological disasters, the inventory (CNR again) of landslides phenomena (IFFI—Inventario dei fenomeni franosi in Italia—Inventory of Landslide Phenomena in Italy), the minutely-detailed study of all the Italian volcanoes (even the underwater ones), data on transport networks and traffic (including the movement of hazardous goods) and weather and climatic data (wind force, atmosphere temperatures and pressures, forecast estimates for hail, lightning, rainfall and snowfall, soil temperatures and storm frequencies).

Despite this plethora of means—and our list has possibly skipped one or two!—many people still perceive the level of tangible results for safety as still unsatisfactory. This is certainly true for Italy and, probably, for many other countries.

Furthermore, in many countries, relevant natural catastrophes insurance or indemnity regulated systems are present, more or less influenced by market-based oriented mechanism or, differently, by State-regulated (or partially regulated) schemes. It is the case of, with different organization, Austria, Belgium, Denmark, France, Germany, Spain or UK (Maccaferri et al. 2012).

Among the reasons for this discontent with all these instruments, we must mention:

- the fact that not all the territory authorities have approved or introduced all the foreseen means;
- some of the authorities have acted only recently—with this lateness resulting effects are yet to be observed and with some territories the prolonged wait has meant conditions have worsened;
- at times, even when they have been applied, the above measures have had limited efficacy because they are seen as simple formalities. Thus, they are not being applied with such clear orders and strict guidelines, resulting in a lack of control, of sanctions, and of cultural participation in the overall aims.

1.4 A Response to the Growing Need for Safety and Risk Awareness: Giving Priority to Information on Disaster Risk Reduction

The complex situation outlined above makes it clear that we need measures to reverse the current situation and its tendencies if we are to achieve safety for human settlements. It is precisely under such circumstances that a new concept of disaster risk for building should emerge today as a new and complementary method of implementation for disaster risk reduction. Such a method will certainly take into account the normative systems referring to environmental, energy and health policies—though more and more often these measures are integrated into systems for guiding market development aiming at obtaining required sustainability. Bearing this in mind, we must point out that the PRED certification is something new that is to be added to existing systems.

It is not always possible to act on the hazard source so as to minimise the results of a calamity—but beforehand, one must take note and act on the "targets", i.e. the buildings and infrastructures, not to speak of the territory context for which we must work out defence strategies making extended use of the possibilities that come from our greater power of risk analysis (cf. Merz 2010).

The data we have collected and given make the size of the problem clear and show the urgent need of action that would at last be effective. To obtain best results it is absolutely essential to involve all the "consumer citizen" who must be sensitised and active about one's own safety—so the individual might refresh, from his origin and instinct, that natural earmarking of investment in his own physical survival (Mumford 1961). We have lived beyond the historical times of fatalism and not-knowing, and those of scarcity. Today's society has the means to face and deal with situations which earlier were taken passively for granted—though in reality they were never accepted as normal.

Suffering for calamity effects is nowadays much more intense since the population is unprepared to take part in what has been called *risk society* (Beck 2000). The damage calculated after each calamity is again and again defined as "huge" or "devastating". The disasters are termed "unforeseeable" or "unexpected". Faced with such news, an individual has hardly any bases on which to act (first) or react (later) both materially and psychologically. Such a situation leads us to insist on the need to make a precise effort to give priority in the management of built-up areas, to the reduction of disaster risks. And this means getting the consumer-citizen as owner on participating in the building's decision-making, to possess more interest and commitment in the building's protective function in both physical and economic ways. There is a cultural step to be introduced and taken. Analysing society willingness-to-pay acceptable level of public expenditure in exchange for reduction in risk, one can note that in many cases the safety demand is higher when the resources are public than, paradoxically, when they are private (Pandey and Nathwani 2004; De Marchi et al. 2001; Fischhoff et al. 1983).

The first steps in this direction can begin only on a solid basis of information and understanding.

To overcome this uninformed situation, this *information asymmetry*, for instance in the house-agent's market, we must refer to the so-called '*house selling information*', which, for instance, has led, in England and Wales (see Part 5 of the Housing Act 2004), to the introduction of the interesting "*Home Information Pack*" (*HIP,* or *Seller's Pack*), that was to be provided before a property could be put on the open market for sale, so as to offer a safeguard to the perspective buyer (the citizen consumer of the building "good").[12] The HIP was suspended in 2010 and abolished in 2012[13] because "expensive and unnecessary" by the new government to encourage sellers to come "back into the market", despite opinions saying that HIP had also selected serious sellers, that had shortened the time taken to sell a property and that probably the related business would survive in another guise (Lisa Bachelor, theguardian.com, Thursday 20 May 2010), contained all legal, administrative and energy information, a set of documents about property such as Energy Performance Certificate, sale statement, local authority and drainage searches, title documents, guarantees etc. And this was particularly useful, in case of leasehold or common-hold.

Other remarkable market-based initiatives, more specialised about risks, are interesting in the frame of our work.

For example the British *Housing Health and Safety Rating System* (Housing Act 2004[14]), which lists 29 types of domestic and civil interior and exterior (excluding natural and man-made disasters' effect) main potential housing-related hazards (for instance hygro-thermal conditions as excess cold or heat, pollutants as asbestos or radiations, food safety, water supply, falling at the entrance or the front door, electric shocks, hot surfaces, explosions and the likes), divides the properties into two main categories (in ten-band classifications) and insists on improvements to be made by the local authority for the lower-category buildings. As a matter of fact, it is a risk-based evaluation tool to help local authorities to identify and protect against potential risks and hazards to health and safety as well as from any deficiencies identified in dwellings—each hazard having a weighting which will help determine how the property is to be rated. It applies to residential properties in England and Wales.

We can also cite, in the USA, the initiative of the "*Institute for Business and Home Safety*" (*IBHS*) in Florida, which is promoted by insurance companies and aims to reduce deaths, injuries, property damage, economic losses and human suffering caused by natural disasters and, as regards other countries, by encouraging good practices and varied systems of *rating* on single risks for new constructions which come up to the Institute's stands. IBHS is convinced that the durability of buildings is a "core societal value", and that with its protection is possible to reduce "human financial losses as well as community disruptions" (www.disastersafety.org).

Furthermore, still in the USA, we can cite, in the framework of the *National Flood Insurance Programme* (*NFIP*), the *Community Rating System* (*CRS*) which

---

[12] HIP became mandatory for homes with four or more bedrooms on 1 August 2007 and was extended to three-bedroom properties from 10 September 2007.

[13] Requirement for the Home Information Pack for property sales was suspended with effect from 21 May 2010 (with the exception of the Energy Performance Certificate, still required). The Localism Act 2011 would formally repeal the Home Information Pack legislation on 15 January 2012.

[14] The building act also introduces a classification for *healthiness* and *safety* (HHSRS), i.e. a method to identify and quantify the occupier's risk in poor domestic surroundings.

is a voluntary incentive programme that recognizes and encourages community floodplain management activities that exceed the minimum NFIP requirements and, as a result, flood insurance premium are discounted on the basis of the community actions in reducing flood damage and encouraging floodplain management. For CRS participating communities, flood insurance premium are discounted in increments of 5 % (from 45 % premium discount in Class 1 communities, to 5 % discount in Class 9). The classification is based on different creditable activities, organized under the following categories: public information, mapping and regulations, flood damage reduction and flood preparedness.

There is also the notable initiative from the car industry, called *European New Car Assessment Programme* (*Euro NCAP*) which gives a realistic independent evaluation of the safety performance of many of the most popular models sold in Europe by means of an efficient rating system, summarising the points scored in tests with a system of 1–5 stars. These initiatives confirm and stress the growing importance of research and development where the work done and its products are summed up and evaluated.

The Predisposition of Resistance of Edifices to Disasters Certification, together with *environment and energy certifications*, may thus mean the completion of an *integrated total information coverage system* contributing to the final definition (from the viewpoint of overall sustainability) of the value of buildings, and not only in a money sense. Such a total definition would stand as adjunct to the just and complete respect of the current norms for building or for maintenance.

1.5 Disaster Certification for Building as a Contribution Toward National Resilience Strategies and Insurance of Natural and Man-Made Disasters

As shown, for instance, by the results obtained with the NFIP CRS and the Euro NCAP initiative, the PRED certification could be a useful step toward the increasing of appropriate resilient strategies and disaster risk insurance, above all, regarding the civil public and private building sector. This is considered "an important issue for citizens, companies and governments across the Union" by the European Commission, as shown in the "Green Paper on the insurance of natural and man-made disasters" of April 2013, and it is a potential right way to correct land use externalities—today in too many countries, not only in the European Union, people do not pay for the risk they generate by living in exposed areas. As already said, this is what generates the violation of the cited intergenerational solidarity.

The Green Paper poses a number of questions concerning the adequacy and availability of suitable disaster insurance in the European Union. Its objective is to raise awareness and to assess whether or not action at EU level could be corrected, or warranted, to improve the market of disaster insurance.

In general, we can say that the final objectives of the PRED certification and of the EC Green Paper are the same—i.e. to promote modern mechanisms as tools of disaster management, to expand the knowledge base in the fight against disasters, and to contribute to a shift towards a general culture of disaster risk reduction.

More specifically, among the Green Paper's questions, many of them are interesting from the potential point of view of the proposal of PRED certification and it is useful to have a more complete picture of its potential role.

To answer the first question "*What is your view on the penetration rate of disaster insurance in the European Union?*" our general opinion refers to hindrances and obstacles exposed in detail in Chap. 2. To list only a few of these, we can cite the most important: unawareness and lack of information on disaster risks, uncertainty of future losses, costs of protection actions, and public re-payment systems. The PRED certification, improving awareness about the situation of each building, could help in reducing the importance of these correlated barriers. More specifically, the bad reputation, be it rightly or wrongly deserved, of insurance companies among consumers (especially common in Italy) is an important additional barrier, on which PRED certification could act thanks to its third-party role which could be useful for creating transparency and confidence to the introduction of a new repayment system after disasters.

To reply to the second question "*What further action could be envisaged in this area? Would mandatory product bundling be an appropriate way to increase insurance cover against disaster risks?*" we can underline that our PRED certification approach matches the technique of bundling together several types of uncorrelated perils into a single insurance policy. The PRED certification, giving a single resistance performance result on the basis of a final global index, renders potentially easier the work of insurance companies (since it could mediate with the corporate strategy in the handling of correlated risks).

Similarly, we can answer about the third question "*Are there any less restrictive ways, other than mandatory product bundling, which could constitute an appropriate way to increase insurance coverage against disaster risks*?" by saying that PRED certification, disseminating awareness and more correct perception of disaster risk levels among different hazards, could be useful, in a framework of transition to law-compulsory insurance system to make it acceptable (even though it is not useful in case of flat rate insurance system). It can be done, mainly by giving a help in considering the level of risk not only for the hazard related to the last disaster described by media or present in the (short) memory of people. As a matter of fact, PRED will be an asset to counteract people feelings towards behavioural phenomena as those causing adverse selection or those related to lack of interest in premiums prohibitively expensive, appealing only to groups of people who are at higher risk and consequently more oriented in purchasing insurances.

Still, in the same way, solutions to the fourth and sixth questions "*How can state or state-mandated disaster (re-)insurance programmes be designed and financed to prevent the problem of moral hazard?*" and "*Could risk-based pricing motivate consumers and insurers to take risk reduction and management measures? Would the impact of risk-based pricing be different if disaster insurance was mandatory? Do insurers in general adequately adjust premiums following the implementation of risk prevention measures?*"—could gain from positive synergies derived from the PRED. Moral hazard (of behaving in riskier ways, having fewer incentives to prevent a loss, knowing that there will be the coverage by insurances) can be counteracted by having people and decision-makers more conscious of the levels of risks of urban assets, which could be allowed by the PRED certification dissemination. Furthermore, the PRED not only facilitates and makes more time-efficient the evaluation of risk levels as well as risk-based pricing policies but it could also be useful in rendering more transparent the process of adequately adjusting premiums after the implementation of prevention, mitigation and protection activities of disaster risk reduction.

As regards data, research and information, the EC Green Paper asks (questions 12–14) how could data and mapping about past and future disasters be improved and "*How could better sharing of data, risk analysis and risk modelling methods be encouraged? Should the available data be made public? Should the EU take action in this area? How can further dialogue between insurance industry and policymakers be encouraged in this area?*". We strongly agree with the statement that "*better information would help to reduce uncertainty*" and with the fact that it makes the relationship between insurer and insured easier along with the economical premium's calculations and the growth of a correct perception and awareness of risks (both at private and public level). Tools as PRED systems, above all in a non-profit or a public-collaborative cooperation framework, allow the availability to the public of all possible data. Researches and technologies of high-resolution mapping of every disasters and hazards are surely a key for PRED's potential success.

More generally, we can obviously agree on the fact that insurance is one of the tools for disaster risk management together with risk prevention, preparedness and response measures, as stated in the EC Green Paper and clearly shown by it. However, it is self-evident that a long path to reach effective and efficient coverage of appropriate disaster-risk insurance and financial risk transfer markets is still in front of us.

In our opinion, a functioning disaster risk insurance system, beyond risk sharing, can be operational at all levels of the risk management cycle, from risk identification and risk modelling to risk transfer and recovery. At the same time, a building's disaster risk certification as the PRED could be useful for and synergic with the insurance's aims (as shown above and anticipated in Bignami 2012). Indeed, insurance plays a specific role but it does not directly prevent the loss of lives or that of other assets, including the historical ones—it helps, above all, to reduce the economic impact and facilitates recovery and reconstruction after disasters. Furthermore, well-designed insurance policies can also work as a "market-based" instrument to discourage risky behaviour and, on the contrary, encourage risk-awareness and actions of "disaster proofing" not only in economic and financial decisions but also, and probably even more importantly, in the land use and urban planning managed by public authorities.

Moreover, we can also imagine that the widespread adoption of the PRED certification use, a sort of third party actor between insurer and insured, could be an important step on the path of the creation an appropriate (and balanced) disaster risk insurance in the EU countries. Promoting knowledge and awareness of building's disaster risk, PRED action could be able to promote safety culture and so to potentially also trigger the disaster insurance-market penetration (obviously, PRED action would be also promoting precautionary actions to risk reduction, as, for instance, public prevention of hazards, private protection and collective mitigation of effects and consequences of events).

To synthesize the whole, we can say that PRED rating system, or the likes, could be an interesting answer to the last question (the number 21) of the EC Green Paper: "*This paper addresses specific aspects related to the prevention and insurance of natural and man-made disasters. Have any important issues been omitted or underrepresented? If so, which?*". As a matter of fact, PRED certification system is a tool very fitting with the EC assertion: "*Risk assessment (including analysis of exposure and vulnerability) is an important and fundamental step in order to inform disaster risk management and the planning process and in order to allocate financial resources. Multi-risk assessments taking into account possible hazards and vulnerability interactions may also help to address correlated risks and knock-on effects. Building resilience is a long-term effort that needs to be integrated in national policies and planning: resilience strategies are also part of the development process and contribute to different long-term policies […]*".

Chapter 2
Choices in Developing the PRED Certification Scheme

2.1 Proactive Approach to Buildings Safety

In Italy, safety initiatives such as the so-called "building certification record-book" (in Italian: *fascicolo di fabbricato*)[1] (cf. Consiglio di Stato 2007)[2] have been put off for an indefinite time: the "file" contained procedures compulsory for all public and private buildings (Cattanei et al. 2002). Thus, the improvement in safety check systems for already existing buildings has remained unapplied despite the needs of safety checks. Disasters come to mind—not only the three latest Italian earthquakes (Molise 2002; L'Aquila 2009; Emilia 2012) or those in Japan (2011), New Zealand (2011) and Chile (2010), but also three recent extreme Italian cases in the Naples

---

[1] This was the result of two serious incidents: in Rome 16th December 1998, the collapse of a 5-storey residential building built in 1952–1954 (27 deaths); in Foggia, 11th November 1999, the collapse of another residential building, 6-storied, built between 1968 and 1971 (67 deaths).

[2] With Ordinance n° 1580, 27th March 2007, the Italian Council of State (Consiglio di Stato) gave its final decision on the "building certification record-book" set up by the Lazio Region and the Rome city council and already declared illegitimate by the Lazio TAR (Tribunale Amministrativo Regionale) with its sentence, 13th November 2006. The file had already become law (n° 31, 2002 by the Lazio Region). This act enabled local councils to set up a file on each building so as to define its conservation and maintenance condition and also its eventual risk situation. The Rome city council adopted the act (n° 27, 24th February 2004) also requiring geological, geotechnical and forestry details for the buildings' land sites. The act also gave the technicians responsible for the inquiry the power to intervene, if necessary, with further inspections, if safety measures were thought necessary. The Confedilizia (one Italian Building Confederation) appealed against the "building certification record-book", criticising, on the one hand, the *onus* on building owners in having to get skilled technicians who would decide whether to call in more experts, and on the other hand imposing on the owner the duty of gathering information concerning planning, urbanism, land registers, structure permits as well as historical, geological and environmental data. According to the judges the building file was "a waste of administration activity" and "an unnecessary doubling of obligations for owners" who "were being forced to supply the City council with information and detail (about their buildings) which have already been or can easily be acquired by public authorities". The latter reference seems to be to geographical-environmental data not usually possessed by owners, but held by specific public offices.

D. F. Bignami, *Towards a Territorial Multi-Disaster Buildings' Resistance Certification*, SpringerBriefs in Environmental Science, DOI: 10.1007/978-88-470-5223-9_2,

area, in Sicily (Favara) and at Barletta,[3] where several people died when a building collapsed because support structures, neglected and worn by time and weather, gave way (not to speak of the Rana Plaza building, collapsed in Savar, Bangladesh in 2013 with 1129 death, and of the last collapse of an apartment building in India at Mumbai in September 2013). In the same way, as already said, in England and Wales HIP was abolished in 2012. The approach shown in the "building certification record-book" and in the HIP initiatives, even though they are characterised by many differences, can be traced back to the logic of the mainly *authoritative* kind found in the laws and administrative powers of the state (cf. Giannini 1979). In this sense, so as logic goes, together with a system where elected representatives decide regulations, a central administration creates the ways for putting them into action and consequently all the citizens and the local administrations toe the line. Such an approach embodying the logic of *command* and *control* has not proved efficient in overseeing the wished-for safety of Italy's enormous wealth of inherited buildings. The reason for this is that the policy was not shared by the majority of the people. There were too many gestures of opposition throughout the countries. The proposed solutions were seen rightly or wrongly, as too burdensome, for instance by some estate agents and building industry associations.

Given the situation, the *Predisposition of Resistance of Edifices to Disasters* certification (PRED or CEPRED) is suggested as a feasible new approach (even though it is still experimental). It would prove useful in the current substantially inert situation, being able to stimulate action in those who want systematise checks on disaster-resistance. This interest spread the culture of buildings' safety. In fact, the CEPRED approach is clearly different from that of the Italian "building certification record-book" or from that of British HIP. Though it attempts to tackle the same problem of safeguarding buildings' safety, it is an initiative encouraging the active involvement of those who are individually concerned with their own building's safety. Such initiative comes from, among others, organisations like *agencies* or *standardisation bodies*, such as the UNI[4] or the Agenzia CasaClima[5]

---

[3] At Favara (Agrigento, Sicily), 23rd January 2010, a small block collapsed, smashing itself to pieces and killing two children. Note that rescuers had great problems as their vehicles were impeded in the very tight street network of the historical centre. At Afragola (Naples), 31st July 2010, another small dwelling, dilapidated outside but rebuilt inside, originally built in the 1940s but restructured in the 1970s, collapsed killing three. At Barletta, 3rd October 2011, in the historical centre, a small 3-storey building collapsed killing five of the workers in the knitwear factory inside the building—despite the many warnings from members of the public about the building's deterioration.

[4] The UNI (Ente Nazionale Italiano di Unificazione, Italian National Agency for Unification) is a private non-profit-making association formed in 1921 and recognised by the Italian State and the European Union. The association studies, develops, approves and publishes voluntary technical norms (the so-called "UNI norms") for all the industrial, commercial and territory sectors excepting electrical and electrotechnical ones. UNI members are companies, professionals individuals, associations, public bodies, research centres and teaching institutes. UNI represents Italy in the European and world standardisation organisations (CEN and ISO, "European Committee for Standardisation" and "International Organisation for Standardisation"). The aim of the UNI is to define technical norms in an effort to improve the efficiency of Italy's social-

("House Climate Agency"), which work at a regional, national or international level. These organizations' aim is to develop and communicate technical norms, which are sometimes voluntary and, at other times, becoming law, represent enforced standards. Such organisations are also involved in promoting research into their relative fields and keeping the public informed.

It is clear that certification can be *compulsory* if it becomes law or *voluntary* when unlegislated but recognised as valid by public or private bodies. In our case, one of the first decisions, taken in the course of discussion to work out forms of certification, was to prefer a working model thought as depending on a voluntary decision by owner or occupier to subject a building to evaluation. In this way, one can take advantage of the benefits the *market* may offer as a result of a more precise definition of the features of the good on offer,[6] features which may also be forthcoming from estate agents.

The author of the present work considers it possible, through organs of "*governance and regulation by information tools*" such as CEPRED certification, to get owners and local administrations to become more aware of the consequences of their choices—regarding the entire field of safety from disaster risk. It may also be possible to promote voluntary action to better the situation by extending the current norms and means of information and awareness. Actually, certification can serve to grasp the limits of possible improvements in single properties. This might encourage intervention as, for example, happens through ecological or energy labelling systems, which are now recognised as information channels thanks to which consumers get to know the ecological characteristics of products (cf. Foglio 2008; Pèrez-Lombard 2008).

At the same time, a proactive approach allowing volunteers to express their opinions can also encourage application and development of certification *in the field* as it has been happening in the provinces of Milan, Lecco and Verbano-Cusio-Ossola (the Italian provinces where we primarily tested the PRED certification). This brings progress in the applicability of measures running *parallel* to essential *theoretical* development necessary for setting a new research project. Our conviction is that in this way one can not only evaluate the effectiveness of the proposal through traditional *experimental* procedures during research but, at the

(Footnote 4 continued)
economic system and contribute to technological innovation—thus promoting competitiveness, trade, consumer protection, environmental care and the quality of products and processes.

[5] Bolzano's "Agenzia CasaClima" ("HouseClimate Agency") is a public body dedicated to certifying the energy systems in buildings. So far the Agency has certified over 2,500 buildings to be found not only in Bolzano Province (in the North East of Italy) but also throughout Italy. The Agency offers training programmes for building workers and promotes initiatives for making the public aware of and sensitive to energy saving, sustainability and climate change.

[6] For instance, when a good's price is fixed or accepted without including all the costs deriving from it and its use, there is an *inefficiency* which may mean excessive production—maximising the net benefit of a private individual or firm and raising unbalanced external costs. Such excesses exist at a remove from the optimum of social equilibrium which could be attained through the free play of market forces (cf. Bignami 2010).

same time produce an improvement in the level of *acceptability*. The latter, of course, embraces awareness of the risk linked to a building's use and, at most, extended initiatives needed for *bettering resistance* to disasters. Such progress should also have a beneficial impact on levels of innovation and on buildings under construction—not to mention profits for the building market (though not due to size and volume).

Only later, if necessary, will we be able to fix new compulsory standards, that is, when the average safety level of buildings has already been improved and advanced in construction—and safety provision has brought costs down. Such developments will, of course, only be possible with a strengthening of building-safety and territory awareness. Such future developments might come about through the evolution of the CEPRED certification itself, which could recover and repeat (in a new positive framework reached gradually) the concept of more or less cogently promoting measures like these in the Italian "building certification record-book", in the British HIP or in the Energy Performance Certificate system, as it is a requirement under EU legislation.

2.1.1 Removing Hindrances to Improvements

Improvement in buildings' resistance to calamities has so far been met with a series of *obstacles*—of various kinds, not all of them financial. Technical and cultural reasons accompany this opposition which sometimes forms a decisive *knowledge barrier*, a wall of diffidence (a true antagonist of the proactive approach).

The decision to improve a building's safety performance is *above all* influenced (differently from energy services, cf. Dall'O et al. 2008; Pérez-Lombard et al. 2008) by the *uncertainty as to whether*, in the period considered, *there will be any disaster at all*, and, in any case, by the *impossibility of knowing when*. As a result, the economic convenience for improvement in structure is not always sure—unless it is distributed over a very long period.

Moreover, it has long been clear that an individual in a similar situation tends to accept the risk—in our case, the risk of an earthquake, an event causing "uncertain"—i.e. unpredictable damage—in an attempt to avoid "losses" considered certain, such expenses and precautions in safety being, too often, considered undue (Kahneman-Tversky 1979). The individual's culture and class may emphasise this tendency which, after all, derives from the individual's income and education (De Marchi et al. 2001).

What is more, as regards "losses", there is yet *another* barrier to consider: in Italy (and not only), in most cases, the *economic losses* due to disasters are covered only partially by the individual and mostly by the *public payment system*, official coverage, and this is collective support based on the general public solidarity. This system's aims run parallel to private insurance-firms and their schemes, except for the fact that the individual bears no direct responsibility—and nor do local or national instituions—in minimising risk; moreover none of these institutions are required to follow rules in deciding when and what payments are made to cover damage.[7] Neither has this system any means for setting aside substantial reserves in order to give prompt coverage when needed. Logically, in this way, political and public financial ex-post interventions by the Governments could have a negative influence on penetration rates both of insurance market and of self-protecting systems. This current system is based on the fact that the damage done always brings human suffering, which leads us to respond responsibly. These situations, for the above reasons, too, *never occur* as one expects, and this is the consumer's economic thinking and approach that runs against any rational consumer's behaviour. Too often the decisions taken favour comparatively cheap short-term solutions—but more expensive in the long-run compared to other cheaper long-term solutions. This is true for one deciding individually (as likelihood) as it is for society as a whole (since floods and earthquakes concern an entire nation, affecting, as they often do, large areas of a country's territory). Thus, solutions are adopted that cost less for the individual decision-maker in his immediate future, e.g. solutions not investing in risk reduction. And this pushes up the overall cost for the nation since damage has been done and has to be paid for.

There is also a delusion which touches the primary function of buildings, an illusion that creates yet another *third* barrier. It concerns the "*inadequacy of public economic incentives*", which are essential and strategic for a building's function. Italian Government incentives for renovation, also for reasons concerning public finances, are not diversified according to the type of work foreseen (with the important exception of energy efficiency). These incentives, then, do nothing to lead the consumer towards a choice to ensure a building's safety. Paradoxically, some of the modification choices made by the consumers reduce the building's safety level—like making structures higher, knocking out walls and so on. There is also the contradiction that life-assurance companies (whose intervention is after death rather than trying to preserve life) enjoy tax incentives—though, of course, these and other insurance firms make a great contribution to social well-being and increasing community resilience.

For Italy, in particular, we must point out a *fourth* barrier to improving buildings' resistance to disaster damage. This obstacle lies in the *scarcity of checks carried out on the observance of regulations*—a scarcity that comes short of many real needs and must be put down to a failure in individual responsibility, which should come from law-abiding habits and professional ethics. But quite a number of decrees and laws have, after the event, accepted and confirmed illegal procedures ("*condoni edilizi*" amnesties for infringement of building regulations) or have diluted decisions favouring safety (e.g. cancelling or *declassifying* local areas from the lists of seismic zones or *blocking the demolition* of illegally-built structures).

---

[7] See in Italy, for instance, the recent floods in areas such as Liguria, Veneto, Cinque Terre, Lunigiana and Genova or see the earthquakes in L'Aquila and Emilia. Such events have led to a continuing debate on the time and costs to cover the damage done and have produced long-lasting negotiations to find the finance by means of new taxes or various allowances in budgets.

Another, *fifth*, barrier against improvement is likely to be *overestimating intervention costs*, which, as is also true for energy-consumption reduction (cf. Dall'O et al. 2008), are taken to be higher than they really are, so that *a priori* the idea of intervening on existing buildings or putting innovative solutions into new structures is abandoned. Furthermore, this overestimation is often coupled with an additional *underestimation* (also called risk *myopia*) of the real costs linked to a disaster risk.

The mental obstacle noted above is often accompanied by a *sixth*, i.e. the lack of up-to-date information and training on the part of many construction professionals. The latter, confident in their experience over years of work, tend to adopt solutions they have already used so as to avoid any complications that new approaches may bring—including the worker's new training, suppliers' non-cooperation, etc. Such professionals also prefer to avoid retraining and having to familiarise themselves with new technologies, such as the latest software apps.

In a similar way, a *further barrier* is represented by the property owners *lack of information*. Such unawareness brings a stalemate to any idea of requesting necessary interventions. Most owners are not able to weigh up accurately the need for intervention or the measures needed to improve safety. Moreover, their approach to disasters is often too *fatalistic*.

And finally, the *eighth* obstacle to contend with is the weakness of the policies that should work in favour of study and *research* on land-and-construction-safety. Safety could be furthered by special projects or advanced schemes for *investigation, industrial research, experimental or "precompetitive" development* and by *informing* the public of results and solutions. This eighth obstacle of weak policies does not favour innovation in the building sector and such innovation perhaps deserves more attention in the national and international strategies of institutions responsible for collective research policies.

The "*Safe House—Casa Sicura*" (CEPRED) project seems able to help in mitigating the effects of all these obstacles to improving buildings' safety—thanks to its chances of spreading greater awareness and directing greater concentration on its issues. The mass of data acquired and developed during inspections to evaluate each separate building has enabled us to formulate a detailed summary and report (see later section). This achievement means precise information can be communicated widespread on the safety situation of single buildings. It also means an encouraging move towards specific professionally-qualified solutions—as is already the case as regards energy components in the building industry.

2.2 Resistance and Resilience Strategies (Scales and Proactivity)

As is known, *resistance* and *resilience* are the two main "strategies" in ecosystems for survival in accidents or disasters (Pimm 1986). Despite the variability of our work's research-scale, recalling that starting from *regional* investigations we came down to concerns with *single buildings*, PRED's project for defence from disaster risk prizes above all *resistance* evaluation and improvement. Many risk-reduction schemes, e.g. a lot of those lately proposed for developing countries by the UN are mainly focused on *resilience* (used as *umbrella concept*).

Here we must specify that by "*resistance*" of a building we refer to its strength or capacity not to suffer from any calamity—until the latter reaches certain intensity—or, more simply, the quality to resist damage or destruction from disaster phenomena.

On the other side, *resilience* concept is the subject of much research and interpretation. But, in our thesis, it is reasonably taken to be the quality to absorb a change, an effect caused by a phenomenon of a given intensity, i.e. the ability to recuperate after damage or destruction and persist. In other words, it is the ability of a system to return to a state of equilibrium (*stability*) after a temporary disturbance (Holling 1973). As a result, with particular reference to natural hazards, resilience could be defined as the capacity of a community of organising itself to recover from a disaster. Nonetheless defence strategies based on resistance and resilience are not equivalent though both work in the opposite direction to vulnerability (Timmerman 1981; Dauphiné 2001; Klein et al. 2003). Strategies based on resilience lend themselves to treatment on a regional or city scale (being resilience desirable[8] property also of human systems in the same way as of natural systems)—if combined with warning systems in areas where finance situation allows not only pre-investment in defence measures but also, after the event, fast recovering from the damage done (cf. Bignami 2010), above all if high level of capital-investments have been made and consequently inertia against relocation is strong (sometimes also considering related costs and benefits balances).

Yet, when an owner wishes to know whether his building is safe and if he should set up some defence system, first of all he wants find out whether his building can, for instance, guarantee survival in a calamity (by its building-scale resistance). Only later will he maybe think of investing in an insurance policy so as to increase, if necessary and possible, also the building's resilience (Proverbs-Lamond 2008).

For a community, either a resistance—or a resilience—strategy may be enough to guarantee survival. Resilience, however, implies that in catastrophe we allow for some significant or high losses. And this is clearly not acceptable for the separate members of the community since each of them will independently try to survive or "resist" disaster. Obviously, a community made up of individuals, each increasing his resistance, becomes, as a whole, more resilient—supposing that the resistance levels reached are real and sufficient.

Our option on a resistance approach in facing natural and man-made disasters is also interesting if set up in accordance with the concepts of *adaptation*, *proactive*

[8] Even though, sometimes, if an area is struck by a disaster, going back to the previous state is undesirable, as it would leave the territory just as vulnerable to the next disaster (because the original state was itself vulnerable).

(or *anticipatory*) *adaptation* and *adaptive capacity* emerged in the context of the communication between climate change and natural hazards scientific communities. Since *adaptation* refers to the process of adjustment that takes place in human systems in response to the impacts of climate change and *proactive adaptation* is taking actions before its *undesirable* impacts, PRED certification represents a possible functional way of *anticipatory adaptation* initiative, as an approach based on the increasing of the ability of physical assets to *withstand the impacts* of climate-change and natural hazards consequences. Above all, since *adaptive capacity* is the ability to really facilitate and implement *adaptation options*, but having adaptive capacity does not guarantee that it is used successfully (Klein et al. 2003).

In this framework, if we assume *adaptive capacity* as new shared umbrella concept, we can define PRED initiative as a tool for *proactive resistance* of society in facing disasters, in contrast with the concepts of *reactive* and *proactive resilience*. If a society relying on reactive resilience approaches the future by strengthening the *status quo* and making the present system resistant to change, whereas a society that develops proactive resilience accepts the inevitability of change and tries to create a system that is capable of adapting to new conditions (Dovers and Handmer 1992), *proactive resistance* both strengthens the *status quo* and is capable of adapting to new conditions, choosing the better solution in a flexible manner on the basis of an accurate assessment (as shown in these pages).

2.3 Basic Methods for Developing PRED Certification

Regulatory bodies (Government, agencies, local authorities, etc.) have three main instruments available for encouraging safety in buildings: regulations, inspections and certification. As mentioned, regulations usually establish minimum requirements, in particular in new buildings; inspections are often in defect if compared to the needs. These two instruments, in case of inadequate cultural approach and unclear (from a mass-perception point-of-view) economical advantage, are evidently insufficient to trigger accelerations in the process of existing building disaster risk reduction. In a different way, the building predisposition of resistance to disasters classification approach could become an innovative safety instrument, maybe a more efficient method to protect people, their goods and history.

The classification thought for the PRED certification aims to show, from a safety point-of-view, the *performances* of resistance predisposition in case of any potential disaster that could strike a specific property, according to a precisely defined methodology. On these bases, our proposed PRED certification scheme has been planned, structured and organized as shown in Fig. 2.1.

Both this chapter and the following will describe the above shown column on the developed *Data collection and calculation procedures*, on the *Classification components* chosen and on the *Variable information and conditions* necessary to apply the PRED scheme.

P.R.E.D. CERTIFICATION FRAMEWORK METHODOLOGY		
Data Collection and Calculation Procedures	**Variable Information and Conditions**	**Classification Components**
Audit Process	Disaster Risk Territorial Scenarios	CALIBRATED REFERENCE (or minimum requirements' set up)
Multi-criteria and Specific Risk Analyses	Information on Building Hazard and Vulnerability	RATING SYSTEM
Predisposition of Resistance Indices' Evaluation	Recommendations for Improvement	LABELLING SOLUTIONS

Fig. 2.1 Scope of the proposed PRED Certification scheme

2.4 Audit Process Bases

Auditing (an "audit" verifies by inspection or diagnosis) is essential for any certification that must result from inspection to ascertain or check how far a system or situation meets specific standards. An audit, in fact, aims to evaluate the observance of fixed criteria, standards, fixed objectives (cf. Perotto-Canziani 2008) according to the audit disciplines of *objectivity*, *reproducibility* and *verifiability* (or, some say, *transparency*). This is the reliance of the *evidence-based approach*, meant as the rational method to reach reliable and reproducible audit conclusions in a systematic audit process (in which audit evidence should be verifiable). In general, it is based on samples of the information available and, even though conducted during a finite period of time and with finite resources, samples of the information available are closely related to the confidence that can be placed on the audit conclusions.

Audits can be classified in various formats. The first form is based on the individuals responsible for the audit itself, i.e.:

- persons selected by the supplier (of a good or service) for an internal check (*first-level audit*);
- persons selected by the buyer or user (or somebody at least involved) to check, for instance, a supplier's performance (*second-level audit*);

- persons outside either the above groups, selected according to criteria shared by both the above groups (*third-level audit*).

The second format as a basis for classifying audits refers to the "*target*" to be checked, which may be:

- the performance of an activity (*process-audit*);
- the organisation of a working wide-ranging process of more or less extended parts of a system (*system audit*);
- conformity to final-performance features of a good (*product audit*).

Bearing in mind these formats which are also the forms issued to be filled in (cf. Chiarini 2007), the *auditing* process singled out and developed by CEPRED certification can be taken primarily as a *third-party* check (though it is also possible to imagine a *PRED certification service* to be implemented internally for the assessment of a property portfolio of a real estate company or of an agency managing public assets—in this case PRED could be also taken as *first-level* type of audit). This means checks are thought to be carried out by an independent body—independent of the issue under examination and of anybody is concerned with the results regarding adequate performance (in a specific finished *product* like a building). The PRED checks are requested by the examinees and are recognised by those receiving the good. These checks mainly deal with *one specific product*, a physical object—rather than products manufactured repeatedly and with regards to its specific safety management process. In our case, the request can come from the owner or the administrator of the property. Those concerned with the results of the checks may be the buyers or the occupiers (renters, staff, etc.).

If chosen otherwise, CEPRED could, on the one hand, not have guaranteed essential impartiality or independence in allowing market forces, which it wishes to favour, to function freely, remembering that such forces depend on adequate quality and quantity of information which must be as reliable and trustworthy as possible (so as we consider *independence* the basis for impartiality of the PRED audit and for objectivity of the PRED audit conclusions). On the other hand, PRED certification could not have been able to collect autonomously covered data and details about each of the local sites for separate individual buildings.

2.4.1 A Collaborative Approach for Improving Performances

While impartiality is a feature of this third-party auditing mechanism, CEPRED's proposed system of checks still stands as a means of *improving performance*. And in our case, of course, it is concerned with the resistance performance of *finished* buildings in the context of their specific sites. These models have been thought out and applied so as to facilitate a *collaborative approach* with the evaluators. On this basis, a range of competences and expertise is put at the disposal of the client so as to help in achieving greater awareness of eventual essential moves to heighten the reliability of the building under diagnosis.

It is not the aim of these checks to pass judgements or impose sanctions. In no sense was the PRED certification invented as a substitute for any legal documentation or as a source of information for anyone other than the client unless he authorises it—so that all *privacy* is maintained, both as the law and professional behaviour dictate, as well as confidentiality and security of information has to be assured.

In offering itself as partner or technical consultant, CEPRED innovates what was the system that was proposed by the Lombard Region and Milan Polytechnic, published in the 5th Notebook of Civil Protection: "*Prevention of collapse-risk in residential buildings*". This publication foresaw the realisation, promotion and distribution of a personal *"self-evaluation" card* whose results would be known exclusively to the owner and whose safety standards would not be recognised (or required by consumers). Without essential backup and encouragement, such initiatives, important and worthy as they are, risk succeeding less than expected. In a situation where local councils and public bodies have no resources to spare for such extra activities, related initiatives like CEPRED, relying on visibility and on the market, may have a better chance of success. In other words, the PRED certification is possible to be seen, despite its focus on a specific product, as a process certification partially including considerations to the buildings predisposition resistance to disaster *management system.*

2.4.2 Technical Diagnosis Through Circumstantial Evidence and Documentation

The depth of an *audit* differs a lot depending on the specifics of the research undertaken. It seems logical to suppose that the depth of the analyses depends on their cost and length as well as on efficiency of the auditors. We had to make choices from this point of view, too. In fact, choices occurred to guarantee complete rigour in *method* depth of inquiring to meet expectations, both as ability effectively to influence risk-reduction (efficiency) and to *measure performance-improvement* after modification (cf. Perotto-Canziani 2008).

CEPRED's auditing procedure *does not call for the use of diagnostic instruments nor does it resort to invasive techniques* on structures to define a building's performance. Instead, CEPRED starts from *reconstruction* (historical, too) and *identification* of *possible events* or *negative contexts* that may typify or strike a building since the inquiry is basically analytical (on the basis of a hazard analysis at district level) and also circumstantial. It also goes into the availability and *conformity of documentation*, the building's characteristics and any extra hazards from its *local surroundings*. Thus, the audit goes into the building's situation *vis-à-vis* regulations including *careful and complete visual contact* with the building and eventual measurements and image analysis—if necessary, repeated at intervals. When *doubts* arise requiring instrumental or invasive research, the certification procedure announces this to those concerned and is halted—and no positive evaluation is issued. The owners or clients may take account of the problem, i.e. of the critical situation and make up their mind to proceed with further necessary checks. This is noted in the certification procedure, which *favours* the renewal, and if the new developments are positive (maybe after work to improve safety performance[9]) the checks (and audit) can be completed with an eventual positive result.

In this way, we think we have found a *just compromise* on the extent of inquiries that have to be made, avoiding exaggerated expenses on instrumental diagnosis when there is a *good* chance (though no complete *certainty*) that such diagnosis would be fruitless—but, at the same time, if it appears essential, though the client may not be aware of this, applying deeper instrumental research. Actually, by increasing diagnosis, circumstantial or visual or by identification or by documentation, CEPRED also favours more instrumental or invasive research into buildings.

2.4.3 Validating Reference: Norms ISO 19011:2011, ISO/IEC 17021:2011 and ISO 31000:2009

The international reference norm used as method check for rendering valid the experimental process of CEPRED (or PRED certification or even PRED management system) auditing are norms ISO 19011:2011 "*Guidelines for auditing systems*" (also including internal or first-party audit, possible for the PRED) and ISO/IEC 17021:2011 "*Conformity assessment—Requirements for bodies providing audit and certification of management systems*" (focused on third-party conformity assessment activity, as mainly possible for the PRED). ISO 19011:2011 *international standard* is a second edition that cancels and replaces the first edition (ISO 19011:2002[10]), which has been technically revised. In particular, its scope has been broadened from the auditing of quality and environmental management systems to the auditing of any management systems. However, as explained in the introduction[11] of ISO 19011:2002, it could already able to be applied to other types of audit as long as professional competence in the matter has been assured as regards environmental context. As for the makeup of audit groups see our following sections. Furthermore, ISO 19011:2011 clarifies the relationship between ISO 19011

---

9 By means of *preventive* or *corrective* maintenance—*preventive* in repairing deterioration due ageing structure or plants; *corrective* in replacing obsolescent features which may be economic, technological, functional or normative in nature.

10 In Italy UNI EN ISO 19011:2003, on the basis of which CEPRED was developed.

11 From the text of ISO 19011:2002 "*ISO 19011:2002 provides guidance on the principles of auditing, managing audit programmes, conducting quality management system audits and environmental management system audits, as well as guidance on the competence of quality and environmental management system auditors.*

It is applicable to all organizations needing to conduct internal or external audits of quality and/or environmental management systems or to manage an audit programme.

The application of ISO 19011 to other types of audits is possible in principle provided that special consideration is paid to identifying the competence needed by the audit team members in such cases".

and ISO/IEC 17021:2011, which is a second edition, too; it cancels and replaces the first edition of 2006, revising and expanding its scope including requirements for the competence, consistency and impartiality of the audit and certification of management systems of all types (e.g. quality or environmental management systems) and for bodies providing these activities as a third-party role.

The application of such norm is as follows:

- the management of an audit programme (understood as various audits planned to cover a fixed period of time);
- the planning and conducting of an audit of the management system—plans giving details and descriptions of activities and of directions given for managing each audit;
- the competence and evaluation of an auditor and an audit team.

The aim of this norm is to check that fixed requirements are *achieved* and *maintained*. The requirements also fit in well with the features of this particular product, i.e. the finished building whose performance should be guaranteed for many years to come.

A study of this international norms have shown that the CEPRED model, even though, naturally, it is something special (or partial) as *management* system of building safety from the disaster point-of-view, meets with both norms basic requirements.

In reference to what is foreseen in norms ISO 19011:2011 and ISO/IEC 17021:2011 for managing an audit programme, we note that our "reference organisation" is also an area of *territory* so that logic suggests an audit programme is usually characterised by applying an (even large) number of audit plans, considering the number of building (client) to be found on a particular land-area (in practice a special kind of partially grouped multiplicity of *auditees*).

With regards to the field of risk assessment and reduction (or *treatment*), the PRED certification completely meets also with the ISO 31000:2009 "*Risk Management—Principles and Guidelines*" (first edition) basic requirements, that can be applied to a wide range of activities, including, among others, projects, products, services and assets, in case of any type of risk, whatever its nature, obviously including those having negative consequences. First principle of the ISO 31000 is particularly fitting with a potential role of the PRED certification, as a matter of fact, it highlights *risk management* that both *creates and protects value*. Figure 2.2 shows the PRED Certification risk management assessment process on the basis of risk management process described by ISO 31000:2009.

2.5 Specific Certification Coverage

To realize a solid method to express the building's predisposition of resistance, we defined and fixed an overall predisposition of resistance performance index (I_{res}) stated in terms of hazards, vulnerability and other parameters, as showed in the

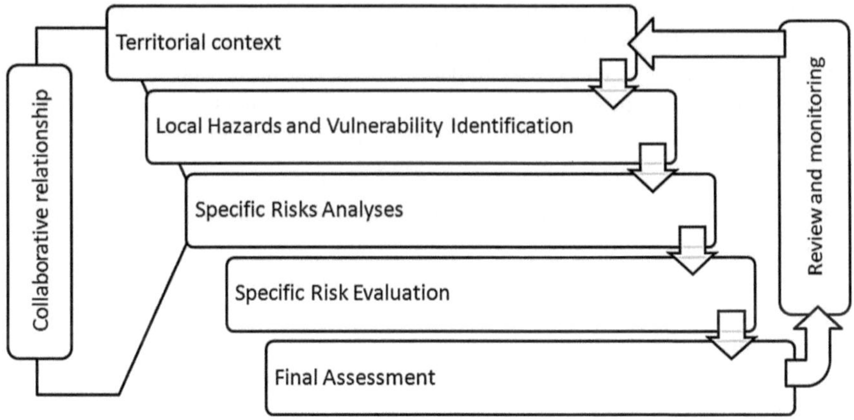

Fig. 2.2 The PRED certification risk management assessment process, in compliance with standard of risk management process described by ISO 31000:2009

following pages, to allow the evaluation of each edifice and the comparison between buildings.

2.5.1 Identification of Disaster Risks for Buildings

As is known, the term "risk" refers to types of event that may be very different from one another. So it was thought necessary and logical to limit the reference range of certification[12]—even beyond simple reference to the already-defined context of construction and territory. This circumscribed reference, in fact, covers the issues of *disaster management* and *disaster-risk-reduction-activities,* corresponding to civil protection as a public function in the sense that these activities have precisely defined common traits. This is true if we avoid some misuse and misunderstanding of the concepts of *risk* and *emergency*. The risks analysed for certification possess common well-defined conceptual dimensions, on whose basis they were selected. In this sense, we need to specify better the general definition of risk of ISO 31000:2009 and ISO 19011:2011 as "*effect of uncertainty on objectives*" and "*having whatever positive or negative consequences*". Thus, the dimensions, in our point of view, are: the *involuntary nature* of the events; the phenomena's *time dynamics* (*short* as regards intervention capacity); the *territorial nature* of the events' consequences; *community involvement* (Bignami 2010).

When we speak of the risk as involuntary, or even "*non-intentional*" (cf. Oloruntoba 2013), when the intentional ones have positive consequences are also

[12] Also foreseen in the norms ISO/IEC 17021:2011 and ISO 19011:2011: "*Audit scope: extent and boundaries of an audit*".

called *opportunity* risks and the non-intentional (or *unfavourable*) also *perils* (cf. Borghesi-Gaudenzi 2013), we refer to the fact that the events damaging buildings and considered in certification do not result from individual decision. Thus vandalism is excluded as are the effects of war or terrorism, of burglary or human violence. In other words, the certification refers to, as English has it to "*safety*" rather than "security". "*Safety*", in fact, covers, among other things, defence against natural calamities.

Moreover, the time dynamic of the event analysed for the certification must be such that in a very short spell a building is damaged together with the surrounding ground area. With this shortness an event can be taken as happening in *real time*. We are, thus, referring to a happening in rapid or ultra-rapid time so that there is no time to invent new remedies as it occurs. Remedies must be planned beforehand. The situation is the opposite of one where the negative effects are suffered *later*—so that countermeasures can be taken after the event's beginning. But earthquakes, for instance, have a dynamic by which effects and consequences take place in real time. In contrast, for example, acid rain takes effect on some kinds of built structure over a long period, so logically it does not figure among the certification risks.

Risks are *territorial* in the sense that are "physical", with effects on a building and on lived-in localities or natural land areas with damage to humans or to goods in such areas. Risks to do exclusively with health and hygiene or connected to the well-being individual are already covered by other certifications or disciplines; as an example we can take environmental certifications or those issued by health authorities or agencies for environment protection. Most of these risks are, in any case, not treatable here as they do come with the dynamic of real-time effects. Similarly, the results of crises in the economy, with effects on society and employment, do not touch buildings construction features or their locality for a long or very long time. Such risks are not of the type to be entered into certification.

And finally, the dimension of *community involvement* refers to the fact that, depending on the size of the calamity, anybody in the buildings or nearly may be hit. In an area where one individual has been hit, there must be others who are also damaged. For this reason, small home or workplace incidents are excluded from certification—such problems usually derive from individual behaviour or the use of household electric ware, machines, materials or tools which have nothing to do with a building structure or the physical traits of the area.

The morphology and climate of the Milan, Lecco and Verbano-Cusio-Ossola areas were the first where certification was tried out. This meant we could not only develop our approach theoretically but also we were able to make calibration tests on the fields so as to include some risks deriving from these areas' morphology and climate in our certification. Such themes are a large proportion of the risks foreseen as possible and covered in our certification. Using analogies in causation and method of treatment, the above risks have been split into two main categories—risk sources *outside* (*external*) buildings and those *inside* (*internal*).

For *outside risk* or hazards, selected for effects *not due to human action* and happening *in real time* and damaging a community's land area, territory, the following categories were formed:

(1) *seismic* risk;
(2) *hydraulic* risk: flooding (from river, lake or broken dam);
(3) *hydrogeological* risk: landslides (dry or wet mass movement);
(4) *hydrogeological* risk: avalanches;
(5) *meteorological* risk: divided into snow, ice, storms and lightning, hail and wind;
(6) *industrial-chemical* risk;
(7) *transport* risk: transportation of dangerous goods as well as proximity to transportation routes;
(8) *forest-fire* risk: zones and strips with buildings under threat from nearly forest-fires (interface fires).

Beyond this list, there are some *outside* risks we have not yet developed completely but only as a starting-point theory—for the obvious reason that they are not present in the province of Milan, Lecco or Verbano-Cusio-Ossola and no field testing was feasible. But despite this, they are due to be included in due course in the PRED certification developing programmes. These risks are:

(9) *volcanic* risk (eruption, bradyseism and related phenomena);
(10) *sea-coast* risk (heavy seas, land and cliff erosion, tsunami, floods induced by wind or other seawater floods and even nearness to water ways).

At the moment, the man induced risks deriving from nuclear-energy plants and from disasters linked to the offshore oil and gas industry and transportation,[13] as well as the natural risks caused by insect infestation, drought, animal stampede[14] or epidemic[15] [cf. Guha-Sapir et al. 2012 (CRED)] are not included in the PRED certification.

As regards the risk sources intrinsic to most buildings, i.e. those originating *inside* or internal [including *typical* risks of the edifices' systems cf. Borghesi-Gaudenzi 2013), they have been divided into the following categories:

(1) collapse risk (worn-out structures, instability of sections and portions of the edifices, subsidence, deterioration, etc.);
(2) fire risk (from faults in electric plant, unwise use of heating materials, etc.);
(3) explosion risk (mainly from use of storage of hazardous gases inside buildings).

---

[13] Consideration led to the opinion that no "normal" building can resist or offer significant protection from radiation in the case of a nuclear accident. The idea of signalling distances from the nearest nuclear plants was considered deceptive because of the various factors in the spread of radiation and people's ability to react to it. Factors would be wind direction and intensity (its altitude, too) the amount of radiation released, efficiency of alarm system, etc. In the same way, disasters linked to the offshore oil and gas industry and transportation is yet not included in our rating system (at the moment not even in the sea-coast risk preliminary ideas).

[14] When a group of animals begins to run with no clear heed since it is an act of mass impulse among animal herds.

[15] These kinds of risks are considered, at least in the context of developed countries, not significant as hazard for buildings, or not relevant in occurrence.

With our focus on buildings, the above categories were selected for the effects *not due to human intervention* and happening in *real time*—though much less often these effects hit a *large area* and all its *local community*. These points must be taken in a sense relating to the building size. True, buildings are usually struck in the entirety by events *indoor* and so reference is to the building's occupants and land area. But the effects may also reach adjacent buildings or a *domino* effect can develop involving a chain of buildings—we might recall historical examples, with widespread effects on a city, such as the Great Fire of London in 1666,[16] the Hamburg (Germany) Great fire of 1842, the Great Chicago Fire of 1871, the Great Fire of Toronto (Canada) of 1904, the Bukit Ho Swee Fire (Singapore) of 1961, the great fire in the historic Chiado district of Lisbon (Portugal) in 1988.

2.5.2 Specific Risk Analysis Schemes

As for outside risk, the audit scheme adopted was divided in three main phases. These derived from the requirements for calculation procedures quantifying all the various risks. This scheme also refers to the simplified expression of risk accepted by UNESCO in 1979 (Fournier D'Albe 1972) and used by any number of authors (cf. Canuti-Casagli 1996; Grünthal et al. 2004; Kron 2005; Bernatik et al. 2013). These phases are appropriate for *risk analysis* and for then putting together the results by a "*multicriteria*" procedure, thought to solve our *multi-hazard disaster risk analysis*.

As is known, the simplified expression of risk,[17] a function of *intensity* I of the disaster and of *element at risk* E is represented by:

$$R(E,I) = H(I) \times V(I,E) \times W(E), \tag{2.1}$$

where H is *hazard*, V is *vulnerability* and W is the *worth exposed*.

---

[16] Starting in the early hours of the night of 2nd September, the fire went on for 3 days, burning much of the city. Some 13,200 dwellings were destroyed, as were dozens of churches, shops and public buildings. A hundred thousand people were left homeless. The fire had started at Thomas Farrior's, baker to Charles II, in Pudding Lane. Accounts of the event say that Farrior went to bed forgetting to put out the fire in his shop furnace; it seems that sparks of burning embers set light to a nearly pile of firewood, hence the flames that spread from the shop to the rest of the city. At that time most of London's buildings were basically wooden structures in themselves reasonably fireproof but these houses were often thatched and thatch can burn easily, ignited by sparks flying in the flame-heated air. The high temperatures reached in the open spaces heated by the blazing rows of buildings meant that materials even those usually resistant could easily catch fire.

[17] Without simplification (applicable only with informed reference to specific *scenarios)*—as in our case, which refers to precise conditions of buildings—the risk expression would refer to *convolution* taking account of the *probability-distribution* of the *chance variables* it depends on.

From this, we get the expression for specific risk (cf. Granger 2003) by the product of the hazard and vulnerability terms above:

$$R_s(E, I) = H(I) \times V(I, E). \qquad (2.2)$$

The specific risk R_s was thought more suitable than risk R in CEPRED's development since it gives the incidence of the damage done by a phenomenon with a specific intensity to the type of "target" considered without including financial or non-financial (extensive) valuations. To include monetary estimates (and ethical, artistic, etc.) and putting "exposed worth" or value into the CEPRED scheme was deemed outside the scope of the intended certification. It is thought that such estimate in economic or value terms (depending on the units chosen to express the value exposed) will have to wait till the buildings' owners, or administrators, request such information—starting from the assumption that each owner wants primarily to know how safe his building is leaving aside its value, the number of occupiers and the works of art it may contain—usually, in fact, such information is already available.

2.5.3 External Risks: Territorial Data Scale and Multiple Criteria Analysis

The crucial methodological key-point we have developed is that of identifying solutions which allow us to classify, compare and jointly evaluate hazards, on which we have scarce if any information available at all, especially about frequency, intensity and, above all, even their special distribution and occurrences including data about high regional variability. Though CEPRED has clearly defined the types of risk covered, the fact remains that the *outside risk* covered in the certification are not always present in the areas where the checked[18] buildings stand. And even when they *are* present, it does not follow that similar types of outside risk possess similar dimensions and frequency. So we need to evaluate such risks on the basis of their importance, i.e. beyond the analyses done to obtain values for hazard and vulnerability. In so doing we can really assign defence priority, instead of doing that on the basis of what would prove to be nothing but an arithmetic average depriving the scheme of effect and function.

This differentiation of territory has required the introduction of a *non-equivalent weighting parameter* (cf. Vismara 2008) for each of the outside risks where the procedure was developed. This weighting *varies* depending on the territorial features in an audit programme. Thus we are able to exploit whatever is made

[18] It would be really useless to enter information telling us that a Milan block of flats is not exposed to volcanic risk as it might be in Naples. Or that a building in Rome faces no avalanche risk as it might in the Valle d'Aosta. Or that a Cagliari school stands no seismic risk as it might in Messina. Such information cannot be used for evaluation or comparison between areas.

possible the aggregation of indices and multiple criteria analysis. This is performed by *dynamic* weighting, mechanically deriving from the area context, and the subsequent normalisation of the results obtained.

We can display the reasons for this choice in a simple but realistic hypothesis together with an empirical statement as an answer to possible objections. If a building turned out to be significantly open to seismic risk (which we take to be the only outside risk present in the examined area), a multi-criteria analysis in which all outside risks had the same weight would produce a largely positive and encouraging result for the building's safety. This would be due to the aggregation of results where the averaged result depends on 7 excellent results out of 8. Such a result could only be misleading as regards the aims of certification. On this basis, one may note that we have consciously chosen to adapt the classical rigorous path of the risk analysis due to the need of comparing risks, finding a hybrid solution. In literature, we have not found a case that included so many hazards in a quantitative risk analysis and assessment over large portions of territory.

If we chose to weight the risk not only on the basis of its magnitude but together with the disaster-frequency or return period, we would be able to proceed unless we worked with arbitrary choices for various risk-types where information, if available at all, is very imprecise or can hardly be hypothesised. This would mean, among other things, introducing greater levels of imprecision (than allowed in our original choices) into the evaluation procedures. If, instead, we went in for more weighting of the outstanding risk in each area, using criteria as objective and shared as possible, we can manage to extrapolate situations and areas taking definite priority for intervention and defence—areas where we must concentrate our operations. If we have partially missed out in the theoretical praxis of calculation, it was in order to gain advantages—on the one hand in getting results that may reduce risks more extensively and significantly while, on the other, the procedure has proved more objective, reproducible and verifiable. In fact, these two goals are generally expected when, and if, at a local government level, a strategy protecting both people and assets (goods) is enforced. Similarly, to follow this objective and match local government strategies, for instance, only with a different solution from an arithmetical point of view, the LEED rating systems makes available some *bonus points* in the category of regional environmental priorities for buildings in different geographic regions (www.usgbc.org/leed).

Anyway, the managing and treating of the hazards and risks in national regulations and professional praxis are similar in many different countries all over the world. One can note an apparently discrepancy between the return period chosen to identify territories to be subjected to strict regulations of land use with the aim to protect people from a potential hazard: usually, in the field of flood, return period are fixed near 10^{-2} or at the maximum at 10^{-3}. Whereas the ways to decide safety distances from hazardous installations start from 10^{-3}, arriving at the minimum to 10^{-5}–10^{-6} or even 10^{-7}, that happen to be not statistically and scientifically representative (Lenoble-Durand 2011). But this is logical, because, despite the famous 1966 flood, river Arno still flows there, when, differently, after the 2011 explosion of AZF in Toulouse, the chemical factories are almost disappeared (cf. Cahen 2006; Bignami 2009). This situation shows that the eventual choice to work on a multi-disaster risks assessment only on the basis of frequencies and vulnerability might be impossible also due to the difference among the generally and socially accepted safety thresholds, because of such huge disparity of orders of magnitudes.

In a different way, the mentioned *Housing Health and Safety Rating System (HHSRS)* is founded on the logical evaluation of both the likelihood of an occurrence that could cause harm and the probable severity of the outcomes of such an occurrence. This ideal combination is possible since it is supported by extensive reviews of the literature and by detailed analyses of statistical data on the impact of housing conditions on health. Moreover, the goal of HHSRS is to carry out a full inspection of dwellings to identify deficiencies of the 29 kinds of hazards considered—which are similar and on which is possible to act both from the point of view of likelihood and severity of damage—in order to determine directly appropriate enforcement action, instead of triggering a mechanism of differentiated-hazard labelling and rating system as done by the PRED.

In the end, the choice of *dynamic weighting* based on analyses of area context allows us to bring out the levels of risk (damage, vulnerability) and priority more clearly. And this can be done following a logical definition of local-area features and on a wide enough scale. Thanks to the dynamic structure of the procedure adopted, we can also evaluate the same buildings several times, choosing, with the collected data, to analyse on a scale for a sub-province, a province, between provinces, or a region. In the same way, we can opt, if necessary, for a *uniform* or *univocal* attribution (uniform returning the arithmetical average) taking a country or continent as a single territory and doing all the weightings on a national or international scale. This option, i.e. of uniform attribution on very wide territorial scales, may seem more suitable for *theory* and *speculation* than to achieve results when it comes to influencing the consumer's specific choices. In fact, the consumer is very likely to make a choice of locality, i.e. where to live in a particular country, region or province. And his choice will depend on *preferences* bound up with other factors like, for instance, the job market, transport, or his personal affections. Only later will he take into account information about the safeness of the buildings he can choose from—one can suppose similar behaviour is even more likely in choosing business or school sites.

These preliminary moves of *initial* examination and singling out *significant criteria* for risks—both approaches linked to territory, environment and city-planning—in as much as these contexts have to be verified for certification (in our case one or more buildings) also have precedents in other certification procedures. See, for instance, in the Norm ISO 14001 and in the EMAS Regulation. In the latter examples, one of the phases of *initial environment analysis* asks for analysis of the context of an organisation by the relative classification in terms of environment and territory, town-planning, and social-economic significance (cf. Perotto-Canziani 2008). Moreover, the ISO 19011:2011 norm itself provides for the audit's field generally to include a description of the physical location of activities to be audited[19] and ISO 31000:2009 starts establishing the context and identifying risks. It is for this reason, in the wake of the above norms provisions, that we generally take an *audit programme* to be a collection of various audits, planned not only over a *precise period* but also regarding a *specific area* which has been decided in advance and fixed in level for each procedure. However, here and in the next chapters, we offer details of a *corrective* inserted into the weighted approach to territory definitions devised, for correcting the potential unbalance that could occur in case of a hazard regarding only a small portion of the overall area chosen as programme reference in relation to the PRED certification. The corrective aims to identify *threshold requisites* for every risk present in the building or the building's area. If such requisites are not met, certification will be blocked regardless of the total points obtained and of the *territory weighting* of each hazard. To define the scale of territorial reference for each risk considered, *specific indicators* (*numerical and measurable*) were used as much as possible. The indicators were taken from the classification most widely used and recognised by modern scientists—such as, among many other sources, there is the *mapping of seismic hazards on national territory* or those areas defined in the *plans for areas at hydrogeological risk*. Our Chap. 4 will contain a detailed account of these choices.

With such indicators as basis, a *repeatable* procedure was applied which, by aggregating *indices*, made the buildings as comparable as possible as well as assuring maximum *objectivity* in calamity weighting and total points.

One good reason that led to this choice was that such comparisons would not have been feasible with the same degree of flexibility and objectivity even if we had used, for example, the Analytic Hierarchy Process (AHP),[20] which had already been applied to similar cases of aggregating risk levels, or the ELECTRE methodology[21] (cf. Deck-Verdel 2012). Moreover, in this way we avoid, for instance, the insurers' praxis when applying the technique of bundling together several types of uncorrelated perils into a single insurance policy; since each peril is independent from any other in the policy, bundling reduces the accumulated risks of any

[19] From the text of the Italian Norm UNI EN 19011:2003 (February 2003)—Note: "*Auditing usually include a description of physical locality, of organising units, of eventual activities and procedures, as well as an indication of the time required*".

[20] The AHP is a method for helping multicriteria decisions. It was developed by T. Saaty in the late 1970s. It foresees, when weighting various indicators/alternatives, the decider's use of pair comparisons and a subsequent hierarchy recomposition of these. Its application requires numerous interviews which, in our case, should be made specifically for each activity programme. What is more, the eventual results from different areas would have no comparison value.

[21] Multi-criteria ELECTRE (*Elimination Et Choix Traduisant la Réalité*, in French: Elimination and choice expressing reality) method is a series of specific ways to combine the criteria in a global assessment (here risk assessment) process. The global level of risk does not rely on a mean value computed over the criteria but on the comparison of the studied zones in pairs or with predetermined virtual zones characterising the limits between the classes (being the zones compared with each other on the basis of all criteria). A zone is more risky than another if a majority of criteria (weighted factors taken into account) give a higher risk compared with the other.

single hazard in the policy and, in doing so, it reduces the costs. If this is logical in a selling strategy, "probabilistic underestimation" is not fitting with PRED rating system objectives. Our work is more a "half-step" towards the *combination* of the occurrence of several hazards rather than to taking into account the possibility of each disaster separately.

However, our approach allows for information synthesis deriving from numerous variables and based on the introduction of an *Index of edifice's resistance predisposition to external risks* $I_{res(ext)}$. This index is a function of the different levels of building's resistance to specific risk $R_{s(i)}$ and of an index of generic vulnerability, V_G (to be illustrated in our next chapter). It can be obtained as:

$$I_{res(ext)}(R_{s(i)}) = V_G \times \sum_{i=1}^{n} R_{s(i)} \times p_i,$$

Here, since $R_{s(i)}$ *is a function of* H_i *and* V_i*, according to the following, as taken from* (2.2), *we obtain:*

$$R_{s(i)} = H_i \times V_i.$$

Inserting this into the previous equation, we can get a total final form for $I_{res(ext)}$:

$$I_{res(ext)}(V_G, R_{s(i)}) = V_G \times \sum_{i=1}^{n} H_i \times V_i \times p_i, \quad (2.3)$$

where i stands for the various n risks in the areas (with, in our case, at the moment, the number of outside risks $n \leq 8$, i.e. the number of *outside risks* possibly present in the area where the certification procedure has been applied); p_i indicates the weightings for each of the n risks (with $\sum p_i = 1$, where values for p_i are obtained after the necessary *normalisation*). Note that for CEPRED, the values of H_i in particular will not be likelihoods (measure unit t^{-1}) but will represent an index of the hazard level, the H_{iL} non-dimensional.

These values are temporarily valid for certification as long as the area conditions do not provide replies that nullify the entire outcome of the certification (answering to a so-called "*red query*", on this see Chap. 3).

2.5.4 Weighting and Compatibility

Formula (2.3) above allows us to progress from the analysis of individual *specific risks* for each *hazard* considered, to a single value, which means we can compare different buildings in the same area while taking the disaster *intensity* of the area into account. As hypothesis, if we divide the areas under analysis for auditing into *zones* with equivalent intensity j for each risk i considered, following upon a procedure to discrete these values' reference-scales, we obtain a basis formula useful for weighting the "ith" risk in the area—useful, too, for being repeatable in the various areas. The formula is:

$$p_i = \frac{\sum_{j=1}^{n} S_{ij} \times Cm_{ij}}{S_{tot}} \times C_{HS(i)}, \tag{2.4}$$

where S_{ij} is the *surface* of the examined area under the risk "ith", risk with intensity-class "jth". Cm_{ij} is the *coefficient of magnitude* for the ith risk having intensity-class "jth". The $C_{HS(i)}$ coefficient (or *past-incidence coefficients*) is intended, depending on the hazards, to take account of the area's *historical* and *social* tendencies, which are very often fundamental in bringing out the more or less marked presence of a hazard and have not yet been included in the used or usable conventional magnitude-scale. S_{tot}, instead, is *the area surface in question (or area scale)*, for which ∀ i is valid:

$$\sum_{j=1}^{n} S_{ij} = S_{tot}.$$

An *area scale of reference* was set up taking in a province's entire territory—as regards the province of Milan, Lecco and Verbano-Cusio-Ossola. We also could cover areas that were more or less inter-municipal if only more homogeneous from a hazard viewpoint. As for the risks in the territory examined, areas were identified with similar-hazard levels through subdivision into *hazard-classes*.

With few variations for special cases, these hazard-classes attribute magnitude-coefficient values Cm_{ij}, variable from 1 to 5. Such values are *provisional* considering the experimental nature of the proposed certification—though, in fact, they were obtained through specific *sensitivity analysis*. These values are still being subjected to research and further verification since broader sets of data are needed for their calibration and validation (see Chap. 3). Obviously, weighting obtained in this way needs its values to undergo a final normalisation—in line with normal praxis and compatibility.

The advantage of working on a new scale for territory that is widespread like that of a province/district (or even several provinces/district[22] or an entire region/state, depending on the administrative country's organisation) is that one can guide a consumer to choose safer areas and already urbanised stretches with higher standards of disaster resistance. At the moment this option is also supported by the available data and mapping. The data are very often organised by province or region's administrations. Much less often, it is practised homogeneously on a smaller scale. In any case, it allows homogeneous analysis of smaller areas, though without any gain in precision or topographical definition in area details. On the

22 The average surface of the Italian provinces' area is ca. 2.700 km^2.

other hand, taking on areas that are too vast may mean losing outstanding territory-connotations and important risk-features. Whereas selecting limited country areas may fit in with good local-safety-strategies. This is particularly true when dealing with areas inside a province (or shared between provinces if the data are homogeneous enough to allow it). Such areas are especially differentiated in some way or are characterised by morphology or environment—think, for instance, of small island zones or the Prealpine contexts, etc.

The CEPRED scheme is different from some typical procedures for energy or environment certification, schemes intended for global application and the inclusion of vast areas and stretches of territory. In contrast, CEPRED can work very flexibly and is *specially suited to the intervention scale for the risk-reducing strategies* as it usually, and justly, happens in case of a fight against natural and man-made disasters. With this adaptability, it can, from time to time, be used by municipal or country authorities. This is, for instance, possible in Italy, on the basis of what is foreseen in the programmes of foresight and prevention that provinces, regions and Government have to realise under law 225/1992 and the Legislative. Decree112/1998. In the same way, it could be, evidently, possible in other countries, for instance in France, see in this regard the *Plan de Prévention des Risques* (under law 565/1997).

2.5.5 Internal Risks: Through-Thought Check-List

Unlike the external risks covered in the certification, the same internal ones are present in practically every building in the territories considered. The rare cases where there is an internal diversification of risk occur when the building have no gas or fuel supply. But this possibility depends on what use a building is put to—the reasons are neither geographical nor territorial. The above explains why, for internal risks, we have decided not to use a territorial reference scale for risk-weighting.

The *Index of edifice's resistance predisposition to internal risks* $I_{res(int)}$ derives from the sum of *points* awarded to risk in compiling a evaluation tool *check-list* where the total points given depends on a risk's importance as recognised in the general literature on the subject (for coverage of this, see the following chapters). Different approaches based on points awarded allow us to synthesise information obtained from many variables so that we get to the *level of specific internal risk* $R_{int(k)}$ for each internal hazard k together with the *vulnerability of the context* V_C. From these we can obtain $I_{res(int)}$ following the formula:

$$I_{res(int)} = V_C \times \left(\sum_{i=1}^{k} R_{int(k)} \right) \Big/ k, \tag{2.5}$$

where $R_{int(k)}$ stand for the sum of the points in the check-list inquiries and responses on internal risks. But in this case, too, the values are only valid for certification on condition that a so-called *red-query* reply does not annul the entire certification (as mentioned, see Chap. 3 to the complete explanation about it).

2.5.6 Combination of Internal and External Risks

So the *Index of edifice's resistance predisposition to disasters* can be obtained by combining indices (2.3) and (2.5) i.e. the function:

$$I_{res} = f\left[I_{res(int)}\left(V_C, R_{int(k)}\right), I_{res(ext)}\left(V_G, R_{s(i)}\right)\right], \tag{2.6}$$

which represents a normalisation by which both indices have equivalent weighting—this goes for our experience so far though, in future, one may give each index a different weighting.

2.5.7 Reliability of Basic Sources for Certification

It was realised during experiments that, in each reference territory, hazard (and related weights for the multicriteria analysis) cannot be estimated with the same precision for each activated programme in different provinces/districts. And, logically, this also applies in part to the buildings' vulnerability since uncertainty levels are heightened. The reasons for the imprecision over territories are varied and mainly regard the completeness or availability of data or of management tools for territorial planning. As a consequence, we decided to introduce a percentage *reliability*-index A_f (in Italian *affidabilità*) on the completeness of information and sources—particularly as regards territory levels of hazard. This index also serves for building performance valuation. It synthesises and shows the upper and lower limits' degrees of *reliability* and credibility. Low values in this source reliability index A_f imply deeper uncertainty for the valuation results in certification. Higher values indicate less uncertainty. Of course, there will always be some level of uncertainty because of the intrinsic stochasticity of calamities.

Creating this index, we think, may serve not only to clarify things for clients but also to stimulate institutions into improving information on their area's risk situation and amplifying true relevant data.

I_{res}, on the one hand, can take the final result of the certified safety-level as the *average value* of a probability distribution, on the other hand, the A_f index can be seen as proportional to the *variance* or the standard deviation of the probability distribution—thus showing the latter's greater or lesser *relative variability* (the lower is the % of A_f, the greater the standard deviation and the possible dispersion of values).

2.6 About the Lexical Option Between Probability and Predisposition

Since one cannot find *return periods* or *frequencies* for every kind of calamity considered in the PRED, the *Index of edifice's resistance predisposition to external risks* $I_{res(ext)}$, just like the Index for the internal risks $I_{res(int)}$, which will be detailed in our next chapter, need to be converted into a unified scale obtained from several different unit measurements.

As we remarked before, for some natural phenomena, one can at times refer to a *frequency-magnitude* relationship such as seismic or hydraulic risk, at least with some likelihood but within all the known limitations. Yet, despite all our efforts, it is still hardly possible to determinate these relationships for other phenomena, such as, for example landslides, avalanches, forest-fires, industrial accidents or building collapse, because of the outstanding influence of human factors or of particular features of the hazards themselves. As for industrial accidents and building collapse, one can make only partial reference to the reliability or trustworthiness of some components. Yet, another problematic phenomenon is the explosion risk from gas leaks.

The calculation of hazard (as well as, obviously, vulnerability) is made by resorting to the relevant *indices of hazard* (and of *vulnerability*), which had been worked out for each of the 8 external risks. Moreover, calculation of specific risk for each of the 3 internal risks was done according to an *index of specific risk*. These calculations are developed from the data collected as foreseen in the audit procedure. They lead to *non-dimensional points totals* which allow the various risks to be compared. The above choice, which was inevitable if we were to compare buildings and areas with different risks, also points to our *basic motive* which also led us, within the PRED scheme, to call the certification a document of resistance *predisposition* rather than *probability*, as risk analysts might have expected. Actually, it would not have served our purpose if we had made a distribution of multi-risk probability since variability and chance nature of the phenomena considered would have soon rendered the probability density (or distribution) function estimate meaningless (cf. Kottegoda-Rosso 2008). However, it is clear that resulting indices incorporate some *probabilistic* values, particularly for a few of the risks examined and for the relative values of specific risk $R_{s(i)}$, or semi-probable, at least in an ordinal sense. We should also point out and stress the fact that the applied approximations made by giving non-dimensional indices to all risks, together with their relative values, are more understandable and verifiable as well as more updateable than parameters for the overall estimate of an n-dimension probability-distribution would have been. By the way, in our case, for the moment, $n = 11$, maybe later veering to $n = 13$ at least.[23]

[23] The same choice regarding risk indices has been made in various projects dealing with risk. For example, we can quote the analysis method for the transport risk of hazardous goods in the Italian-Swiss interreg project called "*DESTINATION*"—*Conoscere il trasporto delle merci*

Furthermore, we want here to underline the choice of using the word predisposition has been thoroughly thought-through and consciously fixed. To use the word *resistance* alone, even though associated with the expectations induced by a certification, would mean to present in an incorrect way our work. As illustration goes, it is enough to remind you of the basic concept of the impossibility to reach the goal of *zero risk*, above all because we are working on, and with, safety in the special field of the disaster risk. Despite the application of the greatest effort of risk *control* and of risk *treatment*, there always remains a *residual* risk, not only as a kind of voluntarily *retained* risk but also as "container" of unidentified or worst scenarios of risk. The use of the word predisposition, on the contrary, shows how the investigations of the PRED audit and rating system accomplish a balanced and reasonable effort to obtain and exploit efficiently technical information and data, most of which, partly, already available for consultation, but in a disordered and unfamiliar array, for whom the data could be useful or vital to have at disposal.

Predisposition has also been considered a better option than the use of the word *propensity*. As a matter of fact, predisposition means "to dispose beforehand", whereas propensity is, above all, "a natural inclination or tendency".[24] Thus, choosing Predisposition, we highlight the situation of the buildings which have to be created or prepared precautionally, in opposition to fatalism or indifference.

2.7 Development of Procedures for Auditing (or Diagnosis): Scheme

Auditing procedure stands at the heart of the method for applying CEPRED. But this procedure, as suggested above, does not exist alone. It must be precisely inserted into the procedures for the planned certification. To this end, we developed a special scheme validated on the basis of the ISO 19011:2011, ISO/IEC 17021:2011 and ISO 31000:2009 norms, which, as we have described, were also one of the criteria for judging the methodological completeness of our model.

2.7.1 Scheme Adopted for CEPRED Audit Programmes

As we have shown in our description so far, the CEPRED procedure is more complex than other auditing schemes because not all schemes have to make such

(Footnote 23 continued)
pericolose come strumento di tutela del territorio—Information on the transport of hazardous goods, as a territory safeguard—www.regione.piemonte.it/ambiente/europa/dwd/destination.pdf.

[24] Random House Websters Dictionary. In Italian it is quite the same with the difference between Predisposizione and Propensione.

precise and analytic reference to a territorial scale.[25] Nor do they have to turn to a non-dimensional classification for the various components being evaluated. So the scheme we developed includes some phases (particularly, in the list we give later, phases 2 and 3) implying a particular *initial* development to fix the audit *criteria*. The latter is needed for *activating* the area certification of the building under diagnosis. This scheme, together with phase 1, can be grouped in the *initial definition of territorial hazard*—inherently, as mentioned, with ISO 31000:2009, that starts establishing the context[26] and identifying the risks. This is why, also in accordance with the UNI EN ISO 19011:2011 norm, the first move in certification is to form *programmes* grouping the various diagnosis-requests per territory. The intervention is, at least, partly to agree on and optimise preparatory efforts at activation. The whole scheme, as adopted, is summarised phase by phase in the Fig. 2.3 table.

For this adapted procedure in phases 2, 3 and partly 4, the main reference sources are *maps* and *indicators* already available in the literature on the subject. This explains why the time period is marked as t_f (*fixed time*), i.e. until information and documentation have been collected to complete the audit procedure. In phase 4 (partly), 5, 6, and 7, the applied procedure collects "*raw*" *data*, out of which[27] it goes on to formulate first *data* (analyzed data) that can be used professionally (i.e. *verified* data), and then proceeds to assign *indicators* and then *indices* (just shown). The aim of the process is to arrive at information that is valid as quantity and synthesis, and which can be easily and quickly understood, compared and put to use in advanced certification.

Of course, *indicators* and *indices* must be treated with caution as they give an *empirical model* of reality and not reality itself. Actually, when we move from data sets to indicators and indices, even though both the latter may have been developed and tested thoroughly, we still lose some of the information that the data sets held at the start. On the other hand, this loss is essential in simplifying complexities that in themselves would be useless for making decisions and policies. With this in mind, the CEPRED phases were worked out so as to offer indicators and indices with known standards of reliability and as valid as possible. Thus, the results must prove *understandable* and *useful* to their users and consumers so as to encourage decision-making with a grasp on social aims and ambitions. The results must be *relevant* not only technically but also to declared interests or political concerns. They must, as well, be *highly aggregated*; they may be derived from many components but in the end they must be offered to the non-technical public in a

---

25 From this view point, energy certification in Italy, according to the decree D. Lgs. n. 311/2006, only requires the building's location in one of six climate -zones (A, B, C, D, E, F) so as to signal the various top values for thermal transmittance U, as far as every closing structure is concerned .

26 Establishing the context in ISO 31000 is to pin-point the objective of the organization, the environment in which it pursues those objectives, its stakeholders and the diversity of risk criteria—all of which help reveal and assess the nature and complexity of its risks.

27 Following the scheme of "*The information Pyramid*" (Hammond et al. 1995).

	Phases of PRED audit programmes	
AUDIT PROGRAMME (1 per reference area)	1. ***PRED opening*** and definition of territorial *reference-area* and fixed *reference-period* (choices of the audit *programme* reference scale[a]) – after client/clients formal approval;	INITIAL FRAMEWORK OF TERRITORIAL HAZARD
	2. *Collection* of geographical, disaster hazards and past events data and documentation (of reference-area and reference-period above);	
	3. Initial (*desk*) analyses and evaluation of the extent of territorial hazards (again of reference-area and in reference-period above);	
	4. *Preliminary -stage 1 (desk) analyses of the hazard site* (local scale) of each singular building;	AUDIT STANDARD PLAN (1 per building)
	5. *Preliminary -stage 1 (documental) analyses of hazard and vulnerability of each building* (prior to sending *information form* for documentary collection and preparation);	
	6. *Direct -stage 2* (*field*) *check* on-site and its locality's *hazard* – the check covers all the risks present in the reference -area;	
	7. *Direct -stage 2* (*visual*) *check* on the building's *vulnerability* and on its *internal safety*; *-6.b/7.b (eventual):* phases 6 e 7 prior eventual communication of a detailed *tailored audit plan* for the building whenever its situation is particularly complex[b]);	
	8. (*desk*) *elaboration* of the data collected in the *checklist* and selected from documentation; *8b. (eventual):* request for further checks and investigations;	
	9. ***PRED results*** published (with eventual suggesting for improving the building's resistance performances) and eventual assignment of the *blue-shields plaque* (only in case of positive assessment and under a limited *3-year* period time before the *expiration* or the *renewal* of the certification);	
	10. Eventual ***PRED re-start*** of another audit for *re-checking* of resistance performances (re-certification). The new audit should take place every three years or should be followed-up after any building modifications or steps taken to reduce risk (subject to eventual *revising* of the initial territorial framework).	▲

Fig. 2.3 The phase-process of the PRED Certification audit programmes, in compliance with standard requirements for audit and certification management systems described by ISO/IEC 17021:2011 and by ISO 19011:2011. [a]For instance, the Programme "Casa Sicura 2009—Provincia di Milano"; areas could be alternatively at a sub-provincial (or subdistrectual)or intermunicipal levels, as inter-provincial or sub-regional (or sub-state) levels. [b]Compound buildings (two or more buildings joined) and/or public or private shareholders (various occupiers/owners) as well aspurpose-built edifices employed for different functions etc

reduced form if they are to be understood and useful (Hamond et al. 1995). As a matter of fact, as the reader can verify in the following pages, CEPRED provides not only a final index in case of delivered statement but also a rich and useful collection of other indicators.

Risks management, on the basis of ISO 31000:2009, consists in a process of risk *assessment*, including risk *identification*, risk *analysis* and risk *evaluation*. It is followed by risk *control* and, if necessary, by risk *treatment*, in order to satisfy the established risk *criteria* (based on *internal* and *external* contexts). Furthermore, risk management process of ISO 31000 also includes *monitoring* and *reviewing* risk.

The PRED audit steps from 4 to 7 are, on the basis of hazard and vulnerability component of risk, the preliminary documentation and on site *risk identification*, performed by the collection of the *verifiable*, *quantitative* or *qualitative* audit *evidence*. The PRED audit step 8 is the real *risk analysis*.

Step 9 is the risk *evaluation* (audit evidences against audit criteria) for the conclusion of *assessment* (*results*) and for the suggestions for the risk *control* and *treatment*. Its outcome indicates conformity or nonconformity. Step 10 is the *monitoring* and *review*.

The process, as a whole, is performed in very strict connection with the client and based on a precise final communication target in mind, thought to stimulate awareness about disaster risk (see Chap. 3), also beyond the audit conclusions.

2.7.2 Types of Certifiable Buildings

Evaluation procedure has been organised for buildings with comparatively complex functions and organisation and of a size to be considered homogeneous as one unit. Thus, no multi-site audit process is allowed under PRED certification system. Moreover, the audit procedure was developed so as to be applicable to *entire* buildings—and so not to single flats in a block—and eventually also applicable to new-built structures (even though it gives only a provisional finding). The procedure, at the moment, covers each of the following prevalent types of use:

- buildings housing *one, two, three or four families* (typically small villas or small blocks—housing at most for four families but that not needing a condominium overall administration by Italian laws);
- *condominiums* with at least *five families*;
- *office* buildings;
- *school* buildings.

For the moment, we have left out, because of their destination features, their size, their structural traits and their types of management and function: *productive-industrial* buildings (excepting artisan activities), transport buildings like *stations*, *airports* and *logistic centres*; *university sites* and *research centres*; military or police *barracks* (and similarly-used buildings); *commercial centres*, etc. Similar schemes and choices, for example in Germany (Merz et al. 2010) and for important

European projects (Messner 2007), have recently been put into practice though in order to estimate economic damage.

But for condominium or offices, a small number of different commercial uses is allowed (for example, on the ground floor) for retailers or artisans (with storage for a minimum of goods and materials)—such destinations being compatible in use with the safety norms for the entire block.

Next steps probably will be to add other types of edifices as little churches and museums (also starting from the *Damage Assessment Record for Cultural Heritage* in case of post-earthquake emergency of the *Italian Work Group for Tutelage of the Cultural Heritage against Natural Risks* established on the basis of interdepartmental order no. 133 of 23 January 2001 of Italian Department of Civil Protection and the Ministry for Cultural Heritage).

The developed certification model also allows for valuation not only of existing buildings but also *newly-built ones*[28] with eventual rating for *potential or provisional* resistance predisposition, as when construction is finished it will be verified with a final field-inspection leading to the final check and statements of the closing certification.

2.7.3 Audit Apparatus: The Computer-Based "PREDICT—Predisposition of Resistance of Edifices to DIsasters Certification Tool" and the PRED Checklists

In accordance with the praxis of the auditory process, various approaches were devised and used, including those relative to the product evaluation, to define relevant and precise *checklists* for each type of building and for area analysis—for both hazard (above all in area analysis) and vulnerability, especially in building analysis. These checklists are particularly useful in ensuring that the checks are really systematic so as to:

- act as guide for operation during *auditing*;
- act as reference source in the early moves towards documentation, in the following desk analysis and in, eventually, proceeding to final statements;
- help in keeping in mind the original check-data;
- make CEPRED check results and methods unequivocal and repeatable—thus guaranteeing results as independent as possible from each valuation-team.

To this end, the queries and checkpoints were worked out and tested with a view to being as *clear* and understandable as possible and to approaching *measurable* and objective answers.

[28] This possibility is also provided by other certification systems like, for instance, that of the CasaClima Agency.

There is a meaningful sequence in the *work-flow* and the queries. The sequence is shaped to give an analysis following a logical scheme of reference, i.e. from a *macro-* to a *micro*-scale; from documentary responses to field checks (also including talks with those in charge of the buildings); cross checks made on the accuracy of information gathered through the intentional repetitiveness of queries; increased attention to the more frequent *possible criticism* drawn from the best-known incidents and cases in pertinent literature (referring to basic information about types of building, geographical areas, year of construction, etc.); the most logical dynamic of a *visit* to the various parts of a building with the intention of putting the "customer" at ease (from the building's surroundings and to its entrance and its common parts passing through its specific functions and taking in connected and service spaces).

Following a similar logic, for the various locations subject to inspection and checking, we organised:

(a) single audit-forms, divided into sections referring to different *hazards* identified in the reference programme, to internal risks and to the building general conditions and information;
(b) single calculation-sheets for noting and calculating internal-risk indices;
(c) Calculation-sheets for estimating both vulnerability to and hazard of external risks.

At the moment, the single audit-forms to hazards, to internal risk and to the collection of the general information have been made for the Italian typological building context. Obviously, an application in different countries must include a preliminary review phase, based on the materials, the technical solutions and the technology more commonly adopted in the specific building and urban sector of the involved country itself for the application of the PRED.

Specifically, as shown in the scheme adopted, the audit procedure begins with a *desk* analysis of the hazardousness of the buildings' site area so as to identify local hazards with their relative intensities and characteristics. The hazardousness analysis comes after the procedure necessary for defining weightings. After this desk analysis the next step is the *inspection* visit arranged with the building owners or their reps.

At first, the inspection visit, runs through the checklist data (*double copies*, at least). This is a *preliminary* data-check with inspection and technical analysis of documentation which must be made available according to the various legal certifications and authorisations—such as building licence or concession, structure stability, qualifying papers for the structure's present condition, certifications for gas and electric fixtures, etc. This documentation is provided by the owner after he has received an *information prospectus* summing up the requirements for the inspection visit.

Together with the owner or his rep, the visit involves a field check on the *site hazard* concerning the scale of the building. Straight after this, there is the technical-visual direct check on the building's *vulnerability*.

The valuation sheet for the site's hazard grew from 17 pages for Milan Province to 21 for Lecco Province and 22 in their (2012 version) final form. This analysis starts from the previous documentation (if it does exist) or from observations on past events—bringing into play references to mappings and analyses that may be traced in the history archives. Any defensive operations against specific risks are evaluated. Such operations may be thought up after the events of the past and, naturally, only when some defence is feasible are such measures and their maintenance taken into account. When, thanks to monitoring, prediction data were available, certain data were checked—such as, for instance, the speed of a phenomenon's advance or expansion, valued in relation to the level of local emergency-planning. As a principle, when no data were available, for the "predictable" phenomena we entered corrective coefficients with product values leaning to safety.

Later, the eventual type of phenomenon comes under scrutiny so as to foresee *interaction* with buildings and estimate how for loads or pressure can be borne or diverted. Integrating the new risks considered into the developments worked out for Lecco province, the audit pages (checklist) for *vulnerability* in the end totalled 26 pages while, before, the audit pages for Milan-province risks were only 19. Once the inspection visit is finished, the subsequent data progressing takes place with *data input* from the calculation sheets in order to get the *preliminary indices* for inside and outside resistance ($I^*_{res(ext)}$ and $I^*_{res(int)}$) and also to bring out any critical points or the need for more information. Such discoveries may lead to *further checks* or the *suspension* of valuation, while working for yet more specific technical intervention to be carried out on the part of the owner—the positive side is that the owner becomes fully aware of how necessary the work is.

The audit apparatus will be completed with the addition of an instrument called *PREDICT—Predisposition of Resistance of Edifices to DIsasters Certification Tool*. It is a computer-based tool thought to assist in the implementation of the methodology, as required by the different phases of the PRED, and to integrate outputs from geographical information systems and from archives of risk-analyses results at programme level.

Among the obvious and well understandable reasons, the need of an instrument as PREDICT is to assure that our method owns:

- basic accuracy of the complex PRED procedures;
- easiest reproducibility and control or inspection of the overall certification process;
- best automatic assistance of users' skills;
- limitation of excess of subjective sensitivity of professionals;
- supply a reference calculation tool to avoid errors or miscalculations;
- managing of standardized hazards and risks maps;
- objective of making the certification diffusion less expensive in order to limit the impact on purchase or rental prices of buildings.

To develop PREDICT as a useful technical support, the PRED is undergoing a full *iter* of description through detailed process flow algorithms.

Chapter 3
Definition of the PRED Rating and Labelling System

3.1 PRED Verdicts' Range

The scope of the PRED certification is not only to show the results of the predisposition of resistance to disasters evaluation but also to include minimum requirements and labels that allows different users to compare and assess potential buildings selected for a prospective employ.

Once the evaluation process is finished, after the desk procedures and resistance *indices* have all been decided—supposing no further checks are necessary as they have all come to a positive conclusion or critical points have been overcome by intervention or technical investigation- we can precede to the end of the evaluation process. This means the emission of a certificate showing one of the foreseen positive outcomes. On the basis of points obtained, and supposing, as we shall see later, that critical thresholds have been respected, these verdicts may be:

- sufficient;
- discreet;
- good;
- very good;
- excellent.

These verdicts correspond to the award of one to five "blue shields", of the type shown in Fig. 3.1.

The valuation is valid for *three years*[1] before expiring, on *condition* that the area hazard, as ascertained in the audit phases, suffers no change and that there are no outstanding *modifications* to the building or its usage.

If it should be necessary to make *further checks* or to *suspend* evaluation while waiting for the owner's party further specific technical intervention or research, a documentation is given which notes the deficient result in level of the building's overall resistance—with details of each risk. This level corresponds to a *grey shield* intended to stress the difference from the positive result signified by the blue

[1] As the certification cycle, on the basis of ISO 17021.

D. F. Bignami, *Towards a Territorial Multi-Disaster Buildings' Resistance Certification*, 55
SpringerBriefs in Environmental Science, DOI: 10.1007/978-88-470-5223-9_3,

Fig. 3.1 PRED's *blue shield* (*left*), *grey shield* (*centre*) and *blue with red stripe* (*right*)

shield and the fact that the valuation has been *suspended* (grey shield shown in Fig. 3.1).

If the desk-data processing brings out *inadequacies*, i.e. some of the building's disaster-resistance features not curable with limited intervention, the certificate will give the result of *unsatisfactory*, corresponding to a *blue shield with red stripe* (Fig. 3.1). In this case, too, as in the previous ones, the certification carries suggestions for improving the building's safety performance. Figure 3.2 offers a graphic idea of the final potential valuations and the "award" of shields planned by CEPRED.

3.2 The Blue Shield as a Choice of Labelling

The blue shield was chosen as the distinctive sign for CEPRED, so as to be clearly recognisable. A shield is actually a means of *defence* and *protection*, as in the certification process itself. A shield also possesses potency as *symbol* and iconography. As synthesis, the final shield-award, with its different totals depending on a *quantitative scale* for disaster-resistance was considered the most immediate and intuitive[2] means for communicating with clients. Blue shields give a good idea of the disaster-resistance potential of a building as well as suggesting the further "means of defence" that the Safe House project offers through the activities of different national civil protection services.

The "blue shield" hypothesis was tried out during an "*Open Space Technology*" (OST[3]) in an "Environmental Liabilities" international meeting organised by

---

[2] Other graphic signs to symbolise protection or resistance were considered, e.g. a *helmet* (not accepted since too reminiscent of work safety and motorised transport); a *tower* (refused because typical of tourism adverts); a *padlock* (refused as, instinctively, more connected to the security field of activities).

[3] The OST technique is a kind of workshop developed by Harrison Owen. It was especially planned to catch many of the significant conversations which often occur, though informally, as the meeting is closing or during coffee-breaks. For us, it has been a sort of focus group for the choice of the Blue Shield.

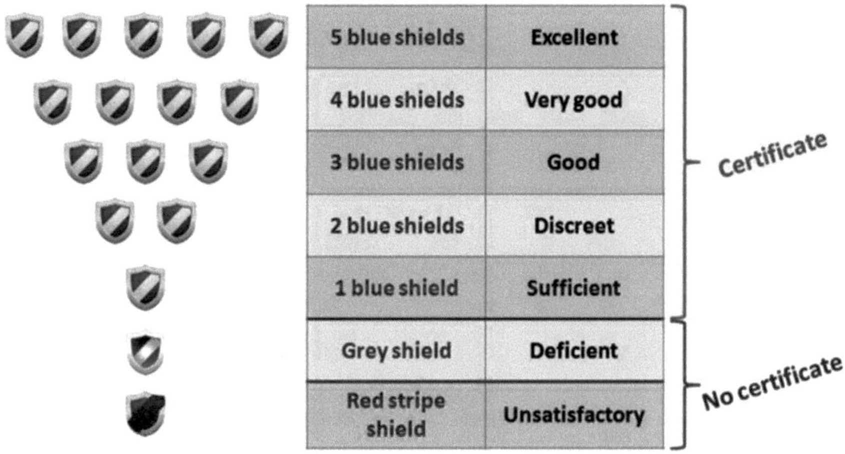

Fig. 3.2 Rating scheme from the PRED classification

INTERLAW,[4] 27th May, 2011. The meeting was attended by real-estate operators from all over the world; particularly those involved in economics and law. After this test was carried out, the "blue shields" have been definitely fixed as symbols used to quantify, in a simplified restitution, the building's resistance under the frame of the PRED rating system.

We can compare our sign system with others that are also successful in communication, recognisability and clarity in expressing quantitative scale—for example: energy certification with classification *letters* and *colours* (as in the EU Directive 2002/91/CE[5] that extended energy certification and labelling to buildings); the EuroNCAP safety certificate for motor vehicles (though aesthetically not very original but always efficient with its *yellow stars*). The latter certificate, thanks to its well-thought-out logo and its quantity-scale expressed by the number of yellow stars (like the PRED scheme with its blue shields), communicates directly and easily with the consumer.[6]

As well as the *certificate* giving details of the valuation, in the positive cases, the building owners are given a plaque bearing in blue shields the valuation of the building's *disaster-resistance predisposition*. Figure 3.3 shows an example of a plaque really awarded to a building in Milan Province in 2009. Figure 3.4 also shows another real plaque for a building in Lecco Province in 2010. Both these

---

[4] *Interlaw* is an international professional non-profit lawyer's organisation.

[5] As changed by the Directives 2010/31/EU and 2010/30/EU.

[6] See http://ec.europa.eu/energy/efficiency/buildings/buildings_en.htm (and, for instance, www.cened.it) and www.euroncap.com.

Fig. 3.3 Plaque really given to a Milan-Province building in 2009 with the level "very good" as level of Predisposition of Resistance. Note that in 2009 the lexical choice was different (safety level, in Italian *Livello di Sicurezza*); since 2010 and 2012 that first choice has changed because considered inaccurate

buildings were evaluated through audits applied and then processed and they were given *Blue-shield plaque* level—"very good" and "good"—with, respectively, four and three blue shields.

3.3 Rating Scale Adopted and Its Calibration

To compare the predisposition of resistance of edifices to disaster, as usual in the case of classifications finalized to release certifications, a benchmarking process should be necessary, starting confronting a building with a proper selected sample.

Fig. 3.4 Plaque really given to a building in Lecco Province in 2010 with "good" result for the level of Predisposition of Resistance (in Italian *Predisposizione alla Resistenza*)—compare lexical adjustment than in the 2009 version

In our case, due to the specific characteristics of the developed indices, the PRED has adopted a *calculated asset* rating system, with respect to the *measured/operational* or the *calculated tailored* ones (cf. Pérez-Lombard 2009). As a matter of fact, the PRED can not be based on a metering on-site and does not depend so strictly, given the building conditions, on everyday occupants behaviour (as in the case of the energy evaluations). Thus, at the moment, due to its start-up phase, PRED rating is implemented with reference values not taken from the building stock but from a standard chosen by us. That's why the PRED is a method for the assessment, under *standard conditions*, of the evaluation of the building resistance performances. However, in the future, our aim is to develop and to hold a database with information of a large and significant number of buildings (categorised by type), so as to permit not only a more complete statistical analysis to improve our initial calibration (for instance, of the mentioned coefficients Cm_{ij}, of the audit

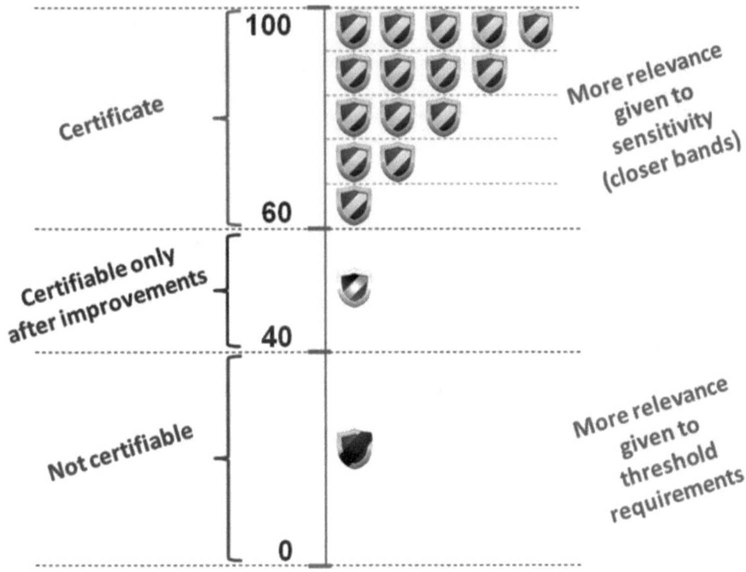

Fig. 3.5 The non-linear scale used for PRED rating system

forms scores or of the thresholds here on the following described) but, eventually, also to obtain a cumulative frequency distribution curve capable to allow us the use of the percentiles as indicators of the buildings predisposition of resistance to disasters.

The *Index of edifice's resistance predisposition to disasters*, based on valuing internal and external risks, is given on the special designed PRED plaque—following on the normalisation of the aggregate results obtained from the multi-criteria valuation of external risks and from the internal-risk checklist expressed in *hundredths* in an intuitive scale. The schematised result is shown in Fig. 3.5. There, *100* is the maximum value, the best possible result. And *60* is the minimum acceptable as satisfactory (the PRED minimum general safety requirement). Thus, the labelled Blue Shields PRED plaque is conferred only if the building has reached this minimum score.

It is important to note that the classes (or labels) to be assigned to the buildings on the basis of the scale that follows, related to the overall index I_{res}, have been designed to be sensitive enough to show the differences among buildings and to confirm the potential future improvements. As a consequence, the buildings PRED comparison, at the moment, is done on the basis of a labelling index illustrating the predisposition of resistance performance percentage (in relation to a self-reference scale).

From 40 to 60 is the deficit zone, where irregularities are considered not so severe and can usually be corrected without too much trouble. This *40–60* band obtains a *grey shield*. In this case, clients obtain only the documentation on the evaluated performance but not the PRED plaque.

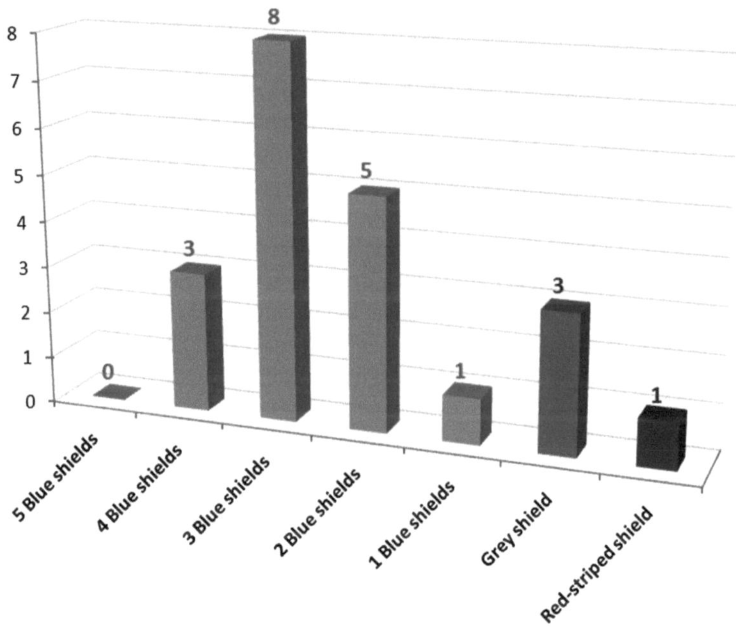

Fig. 3.6 The bi-modal trend of frequency distribution of the first PRED certification results, on the basis of the 2009–2012 calibration campaigns in Milan, Lecco and Verbano-Cusio-Ossola

0–40 is a serious negative result and is marked as *unsatisfactory*, getting the *red-stripped blue shield*. This foresees intervention to improve the performance of the building and/or its landsite. This intervention would be pretty thorough and demanding as regards time, costs and working complexity and the situation might be so bad as to make any intervention out of the question. In this case, too, clients obtain only the documentation on the evaluated performance but not the PRED plaque.

The fixed minimum I_{res} level and the related Blue Shields bands (or intervals) criteria are, as usual in these cases (cf. Pérez-Lombard 2009), subjective and influenced by a kind of "policy decision" aiming at pursuing effective results of disaster risk reduction. In spite of this fact, results obtained during the first *calibration campaigns* in Milano, Lecco and Verbano-Cusio-Ossola offer us an encouraging, even though only at an initial stage of testing, *frequency distribution* covering approximately all the possible results (5 blue shields results were excluded as precautionary "communication" choice, using a special prudential coefficient), as shown in Fig. 3.6.

The frequency distribution shows an interesting *bi-modal* trend, with provisional data *peaking* at both *"3 Blue shield"* and *"Grey shields"* values, because they are not at the extremes but at intermediate values. With a sufficient number of data of assessed edifices (maybe even thousands of them), the *Blue-shields frequency distribution* of a given territory could even become something capable of illustrating an important safety characteristic of built-areas of a region or a province/district. The *Blue-shields frequency distribution* could represent a synthetic *territorial signature*[7] (maybe we could call it *Blue-shields territorial signature*) useful as a new kind of potential aggregated territorial classification to support the civil protection *strategic planning*, in order to improve, from a modern point of view, capital asset and property management of a local community.

The same scale (in hundredths) of I_{res} (Fig. 3.5) is used for the simplified valuation of $I_{res(ext)}$ and $I_{res(int)}$ and the single risks and their shield-definition included on the emitted documentation (or *statement*, in case of a positive evaluation with more than 60). This is done independently of whether the result is positive, in deficit or unsatisfactory but in order to offer, in any case, the client detailed information on each risk present in the reference area of the audit programme.

3.3.1 Balance Between Points Score and Thresholds

International-level certifications or rating tools with a reputation for sustainability, healthiness and safety, generally refer to models that can be divided into two main categories depending on their *evaluation process* of certification *criteria* (in our case we compare *risk analysis results* with our *risk criteria*) (cf. Winward et al. 1998):

- With a *standard minimum threshold* (or *endorsement labels*): the certifiers fix minimum requirements (for obtaining the certificate) through the singling out of one (or two) minimum thresholds. This is done with an essentially *binary* system—of the *on/off* or *conforming/non-conforming* type covering every aspect in the checklists are (it is the case, for instance, of ISO 9000 family standards related to quality management systems or of ISO 14000 standards related to environmental management); thus, there are those that meet the specified criteria and those that do not;
- With a *points system* (or *comparison labels*): the certifiers define a series of factors, which once verified, are grouped into one or more series of scores. The points awarded depend (more or less) proportionally on the performance index values. The final score (total or weighted total) will show whether the minimum for certification has been obtained or one of the foreseen scale-levels has been reached (it is the case, for instance, of Leed rating system or of energy certifications). Thus, in this case labels are in all cases multi-category: all goods, for instance, are attributed a label that classes them from "better" to "worse". Usually this category is more easily adapt to a compulsory use.

[7] Thought of as similar to the *spectral signature*, which can identify univocally an object starting from the specific combination of emitted, reflected or absorbed electromagnetic radiation at varying wavelengths (useful, for instance, in many remote sensing applications).

Instead, the scheme adopted score by PRED classification has a *mixed system.* A positive result is achieved by getting (minimum and above) with some basic and necessary conformity requirements. These must in any case always be respected—independently of the overall score. But the PRED is also a points system (or comparison label) showing buildings that are "better" and "worse" from a *disaster resistance predisposition* performance point of view. This solution was necessarily due to the complexity of aspects of the PRED process, especially regarding the merging of several different hazards. Some theoretical, but possible, examples follow here so as to clarify criteria and mechanisms:

- Even though a building might be perfectly built in a safe area, if it turned out to be an *abusive* structure (i.e. illicit, unauthorised) it would never gain certification. The same goes for buildings without legal final-test documents for their gas, electricity, etc. In our developed checklists, these requirements are defined as "*red queries*" since if they are not respected, certification is blocked and the final shield automatically turns into *red-striped blue shield*;
- After diagnosis and processing, if a building with one or more risks scored not simply less than 60 points but less than 40, maybe because of a single risk earning a red-striped blue shield plaque, then no certificate would be given. And this would be the final result even though the weighted average for other risks compensated the low score (giving temporary scores of 60 or more)—because of the initial weighting of the initial area-risk analysis;
- If two or more risks were valued as deficient (between 40 and 60 points) but the results for other risks were positive, theoretically compensatory, with a 60 or over result, a certificate would again not be awarded and a *grey shield* would mark the deficit;
- If there is only one deficit risk (between 40 and 60) and the results for other risks allow weighted compensation with an overall result of 60 or more, then a *blue shield* is possible, with prescribed interventions to be addressed in an established time.

3.4 Layout of Certification Form

At the end of every building-valuation procedure, the client is given a *certification statement* carrying all the main valuation results—without regard to whether the results were positive, deficient or unsatisfactory (in the latter last cases, it is simply differently called *documentation*). On the first page of the dossier (*synthesised document of certification*), we find the result for the *Index of edifice's resistance predisposition to disasters* I_{res} and the two relevant aggregate indices for the external and internal risks taken together, i.e. *Index of edifice's resistance predisposition to internal risks* $I_{res(int)}$ and *Index of edifice's resistance predisposition to external risks* $I_{res(ext)}$. A propos of this, see the example given in Fig. 3.7 for a building valued in Milan province in 2009. Note that the *resistance indices* in the

CERTIFICAZIONE CASA SICURA 2009

ATTESTATO SINTETICO DI CERTIFICAZIONE

Tipo di edificio	Residenziale Bifamiliare
Indirizzo	Via [redacted]
Comune	Milano
Provincia	MI
Proprietario	[redacted]
Anno di costruzione	1923 - 1929

RILASCIATO DA:
Fondazione Politecnico di Milano
Fondazione Politecnico di Milano
Via Garofalo, 39 - 20133 Milano

CON IL CONTRIBUTO DI:
Provincia di Milano

LIVELLO DI SICUREZZA RILEVATO

$LS_{attribuito}$		Range	$LS_{esterno}$	$LS_{interno}$
	ECCELLENTE	100÷92		
88,2		91÷84	85,9	90,6
		83÷76		
		75÷68		
	SUFFICIENTE	67÷60		
	DEFICITARIO	59÷40		
	INSODDISFACENTE	< 40		

SI ATTESTA CHE IL LIVELLO DI SICUREZZA DELL'EDIFICIO:

☒ E' CONFORME AI REQUISITI ☐ NON E' CONFORME AI REQUISITI

Attestato n.	1	
Data	31/03/2009	TIMBRO
Scadenza	31/03/2012	
Valutato da	Ing. D.F. Bignami	Firma
Approvato da	Prof. C. Baldi	Firma

Rischi valutati: ☒E1 ☒E2 ☒E3 ☒E4 ☒E5 ☒E6 ☐E7 ☐E8 ☐E9 ☒I1 ☒I2 ☒I3

P 1/14

Fig. 3.7 Example of first page of synthesised statement of certification made for a Milan-Province building in 2009

certifying document are, above all, to be clear, given as *safety levels (LS)*. These represent the *100 complement* of the risk indices—high values for *specific risk level* correspond to low values for *safety level* and vice versa.

The document also contains further important information, such as: the data for the inspected building and the *unique progressive* number conferred to the dossier (giving it specific identity also as safeguard against counterfeiting—this identity number is also given to the eventual plaque); dates for issue and validity; fill-in boxes for marks conformity or non-conformity; boxes for the *risks* examined as present in the reference area of the audit programme.

The following pages of every client's statement booklet form a section containing page per page, the *specific statements of the certification*, i.e. results for each risk examined—and researched in the area's audit programme. Our aim is to induce the client to grasp his own specific risk-situation in depth—by using a system that gives clear prominence to details of eventual critical situations. And it does so without bringing the details in index values that are too high. In a synthetic manner, present CEPRED can offer this section in a number of pages which are *specific statements* for an area's external risks (6 pages for Milan, 8 for Lecco and 7 for Verbano-Cusio-Ossola) as well as 3 pages for internal risks. Our Fig. 3.8 shows an example of a specific statement giving the safety level for a single risk—in this case, too, *LS*, the Safety Level, is the 100 complement of the level of specific risk—here *LS*, V_L, H_L and R_{SL} are also indicated (vulnerability level, hazard level, specific risk level). This is considerably significant as additional information so to have an explanation why each risk is important and why it is caused more by the state of conservation of a building than by the type and kind of natural phenomena present in its region. At the moment, no section has been developed to show the weights of different hazards insisting on the PRED programme reference area of the inspected buildings (but further considerations concerning this possibility are taking shape). With Fig. 3.9, we have an example of specific statement for a risk present on the programme-scale but not present on the landsite of the building being valued.

The following section offers summary information about the certification and its meaning. These facts are fundamentally important and so are always included in the client's dossier. The *first page* of this particular section gives precise details of the certification, stressing for the client that:

- the request of certification is *voluntary*;
- in no way does the certification act as a substitute for other documents, permits, certificates foreseen by law or for norms deriving from laws;
- the issuing of the certification and the eventual plaque in no way implies any assumption of responsibility by the certifiers if a calamity should occur[8];
- the certification procedure does not have to check the validity of the documentation offered by the building's owner. If anomalies are found in this

[8] Energy or environment certification also point out *performance*, i.e. behaviour underlying the hypothesis of standard gestures by the occupiers or the builders. For example, if doors and windows are left open, energy consumption will rise above normal. Or, if, when the building ends its useful life and is demolished, the resulting debris is not recycled but cleaned away without differentiation, the initial environment contribution will be partly compromised.

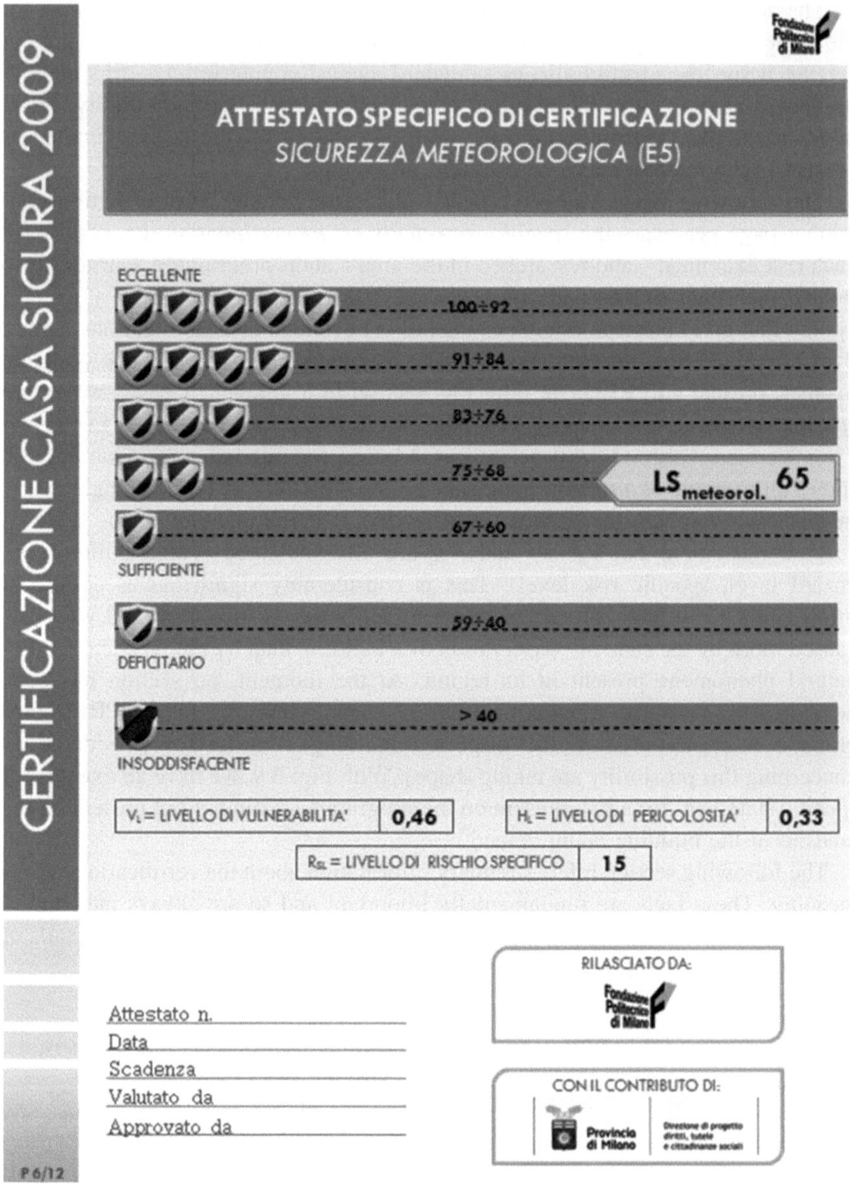

Fig. 3.8 Page of a section with internal or external single specific risk from the synthesis statement of certification for a Milan-Province building in 2009 (with values for *LS*, V_L, H_L and R_{SL} of the *i*-th attributed risk—meteorological)

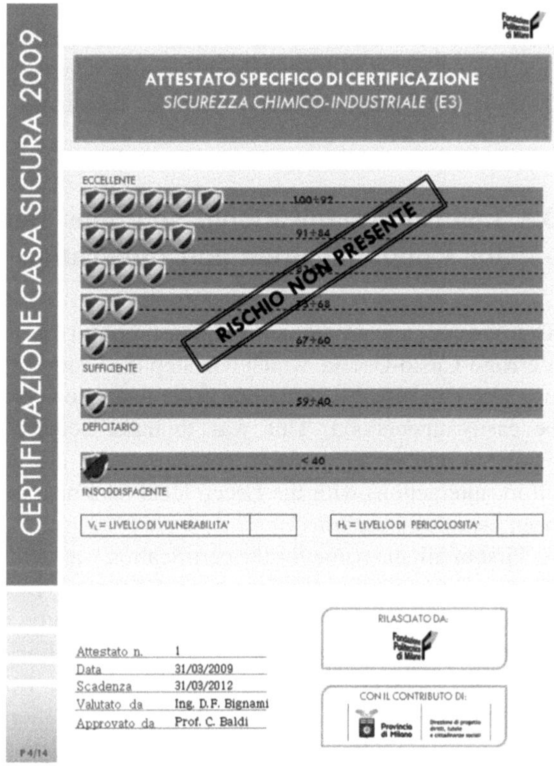

Fig. 3.9 Page of section showing specific single risk in the statement made for a Milan Province building in 2009 (*LS*, V_L, H_L and R_{SL} for *i*-th risk not given since the local scale indicates no risk—chemical-industrial)

documentation, the certifiers will decide, on their own account and following professional practice, whether to accept it as valid;

- the award of the certificate does not, even if the final result is very good, exempt whoever frequents the building from behaving responsibly, correctly and prudently.

The *second page* of this section lists the most ordinary and basic responsibilities and norms in the behaviour of owners, administrators and occupiers as regards a building's safety.

The section's *page three*, despite being the certificate's last page, is one of the most important of the PRED certification. It carries *suggestions for improving and monitoring* resistance and safety. If such ideas have resulted during valuation and processing, they are always given whether the results has been positive or not, i.e. no matter if the building has proved conforming, deficient or unsatisfactory, these suggestions are offered to encourage voluntary action in continually improving each building's safety.

As mentioned, in some cases *prescriptive indications* are inserted into the suggestion for improving safety. The idea is that in order to be sure of obtaining confirmation of the rating level when its 3-year validity expires (or even before this date), these indications should be followed. A concrete example of this type of situation is when there are signs of decay because of, say, wear and tear or corrosion in parts of the building. In such cases, one may say that in the short term there is no problem but if the situation is not dealt with it will worsen in the long run. In such cases, only early warning may encourage the client to act in time and do away with the possible causes of his building's waning safety-level.

3.5 Logical-Organisational Solutions and Concepts for Communication and Illustration

The development of the certification proposal for the provinces of Lecco and of Verbano-Cusio-Ossola, which was step II of our research and development project, was also concerned with finding communication solutions and schemes that could be easily understood. This was to make acquired and processed information available and understandable for those who are not technically expert. In this effort, interactions with the Lecco local administrators and their availability have been especially useful.

First of all, the name for the certification was definitely decided: *Certification of the Predisposition of Resistance of Edifices to Disasters—CEPRED*, i.e. PRED Certification. This version, which was already presented, was preferred to CE.P.R.E.R.D. (Certification of the Predisposition of Resistance of Edifices to Risks of Disasters) because it was thought the word "risk" was redundant for communication and C.R.E.D. (literally Certification of Resistance of Edifices to Disaster-Risks) which was considered incomplete. The choice of name for the certification was made with a number of factors in mind. One was, as mentioned, the importance, for the building's scale, of defence strategies based on *resistance* rather than *resilience*. Another reason was the need to keep in mind the *uncertainty* of the identified solutions which suggest one should be talking of *predisposition* so as to encourage a too reassuring interpretation of the final analyses that is why we do not use, in this context, term such as capability. These, though carried out with extreme precision cannot intrinsically eliminate either the uncertainties linked to disasters or cover all research on eventual unforeseen events—or even on possible inability or professional dishonesty of cowboys in the building industry.

Figure 3.10 shows, following on the name-choice, the proposal for the CEPRED logo to be potentially adopted in Italy (a) and also shows the analogous international version in English (b), in French (c) "*CE.P.R.E.C.—Certification de la Prédisposition de Resistance des Edifices aux Catastrophes*" and in Spanish (d) "*CE.P.R.E.D. Certificatión de la Predisposición de la Resistancia de los Edificios à los Desastres*" [In Italy, since 1/07/2013 Certificazione della Predisposizione di Resistenza degli Edifici ai Disastri and Certificazione della Propensione alla Resistenza degli Edifici ai Disastri, so now they are registered trademarks[9] (at the Italian *Ufficio Brevetti e Marchi*) in a graphic version]. The criteria behind this proposed logo, which must

[9] With the *registration numbers* 0001549270 and 0001549272, respectively.

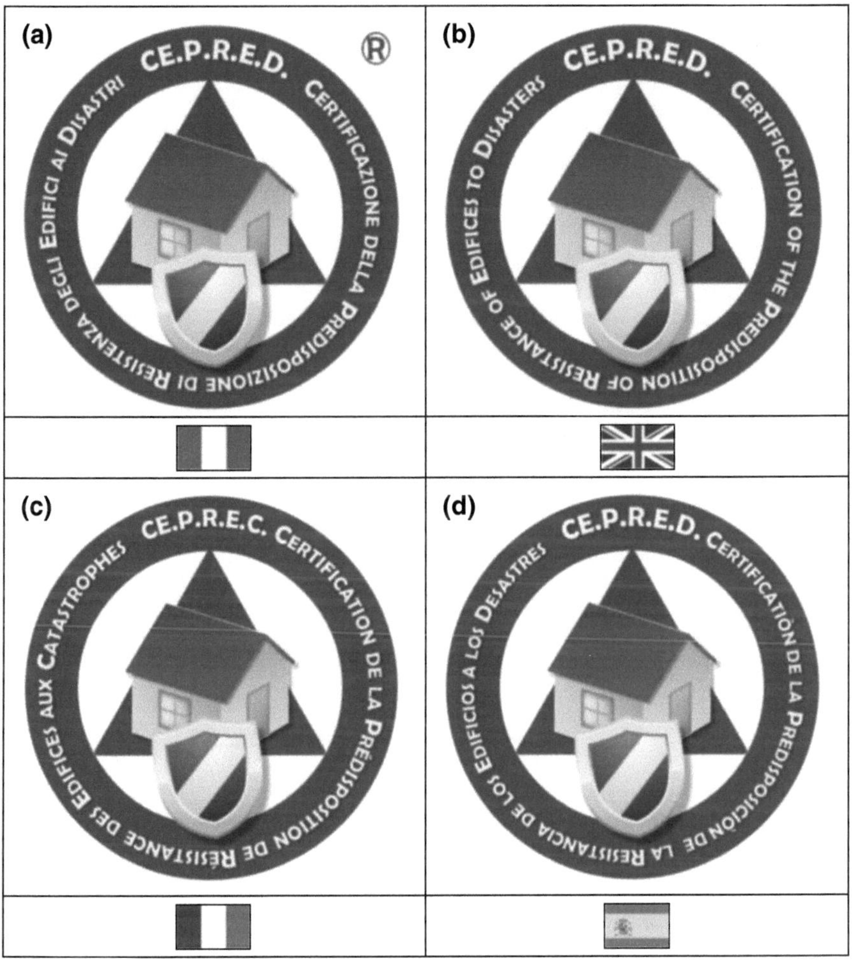

Fig. 3.10 **a** CEPRED Italian logo—*registered trademark* (*Top-left*); **b** CEPRED logo proposal for international use in English (*Top-right*); **c** CEPREC logo proposal for international use in French (*bottom-left*); **d** CEPRED logo proposal for international use in Spanish (*bottom-right*)

ensure immediate recognition and also a grasp of the aims of certification, were as follows:

- a dominance of blue as reference to blue shields awarded to buildings that pass the tests;
- a basic design recalling the logos of local civil-protection services: a circled triangle to represent collaboration and exchange with disaster-risk-reduction teams;
- a building for valuation as centred image: a red-roofed house;
- a blue award-shield placed in front to stress the idea of protection.

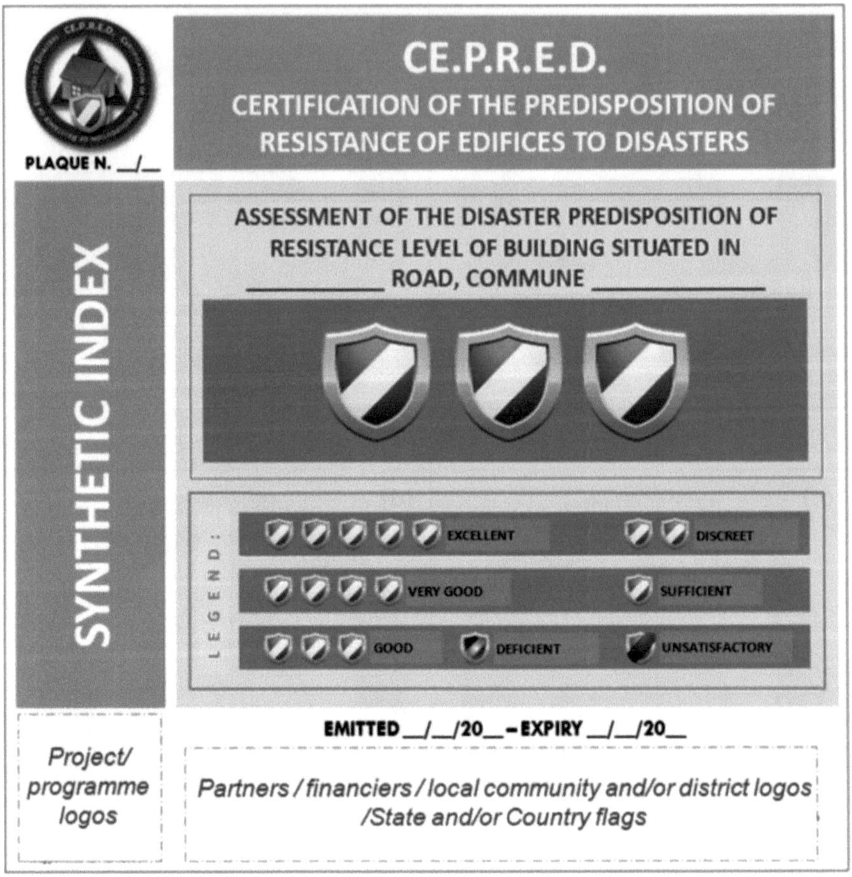

Fig. 3.11 Proposed new layout of CEPRED plaque for international use; note and compare the differences with Figs. 3.3 and 3.4

Once this solution in graphic form was worked out, there were afterthoughts as illustrated in Fig. 3.11 for the certification-plaque layout. In Fig. 3.12, note the space for:

- the rating obtained expressed in shields;
- the progressive numbering of the plaque;
- the data of the valued building;
- the expiring date;
- the spaces for eventual logos of partners, financiers, local communities or provinces/districts or agencies, and, if the case, for instance, EU programme or project (if necessary still at an experimental stage) and flags of States or Countries supporting the audits together with details of the investigated building address.

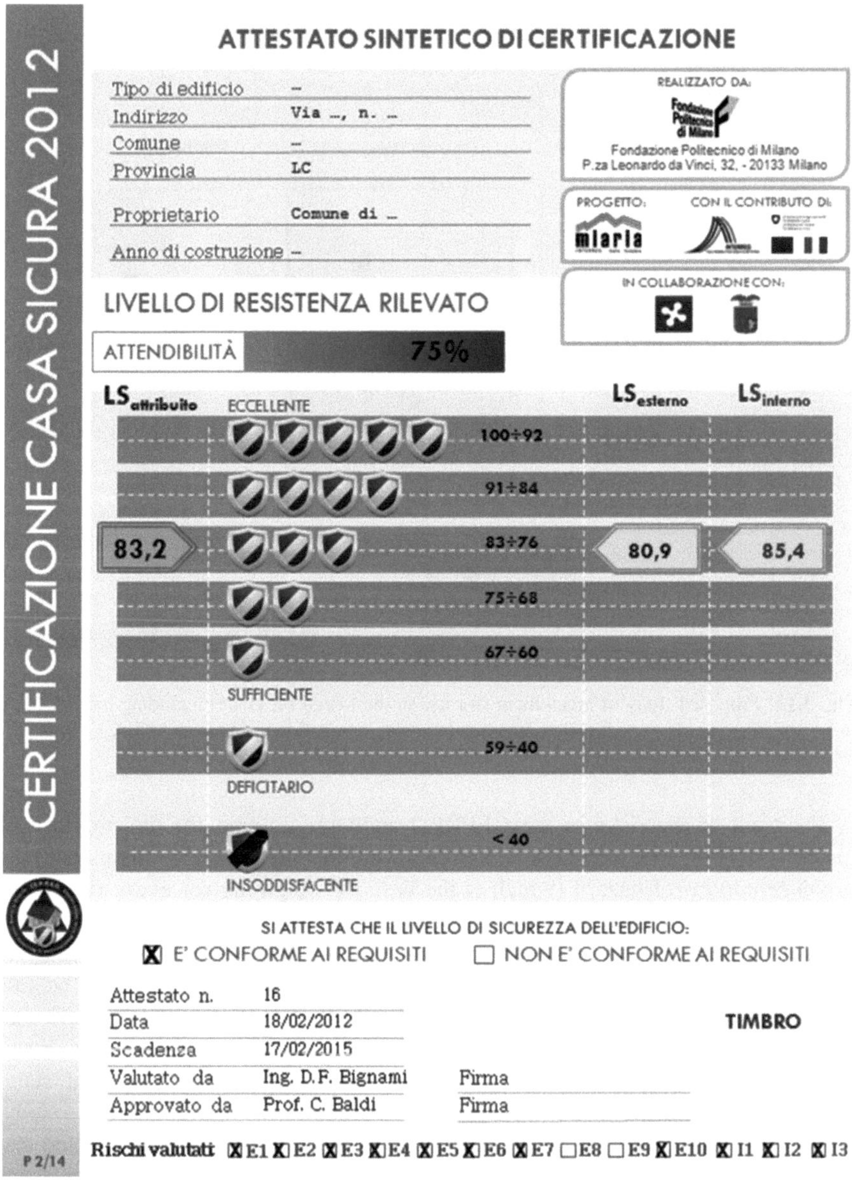

Fig. 3.12 The synthesised statement of PRED certification (for use in the Lecco Province), including reliability index A_f

The latest proposal for the synthesised statement includes, beside the new logo, the new source-reliability A_f index for the data on the territory treated. Figure 3.12 shows the latest version made for Lecco Province, which for this new index obtained a value of 75%, which is a more than discreet reliability.

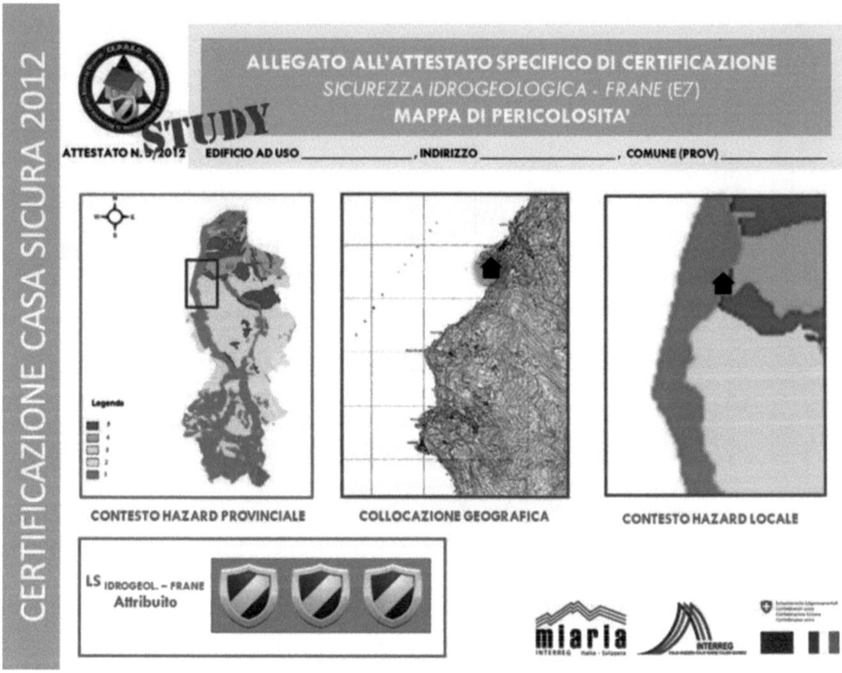

Fig. 3.13 Proposed study of attachment (for use in the Lecco Province) including hazard maps, to be added to specific certification-statements or documentation for various risks

For each of the risks in the CEPRED audit programme for the province of Lecco and Verbano-Cusio-Ossola, systems were invented to show clients both the *province context* of hazard (which is the basis of weighting for every risk in the area) and the *geographical collocation* of the building not to speak of the *local context*. These developments are aimed at aiding understanding and awareness of territory hazard in the client seeking certification. Such a result might be less effective if it depended only upon the award of blue shields, the final synthesis and the number values.

In the case of Lecco Province, an attachment was added to the specific statements of the certification. The attachment was organised as in Fig. 3.13, as a study.

Instead, for Verbano-Cusio-Ossola Province, a study solution was adopted that gave the same information in two different images, one concerning the province context and the other the local context of each risk. These images were both included in the specific certification-statements, in the pages illustrating values of vulnerability, hazard and specific risk levels.

3.5.1 Matrix for Specific Risk in a CEPRED Programme

As already suggested CEPRED takes its place, with promise, in the best line of research and in-depth analysis of calamities and their consequences (for instance, in the emergency planning approach based on the reconstruction of event scenarios). This is thanks to its analysis on a local scale and the details researched. The analysis is functional and derives from contrasting of calamity effects whose data can be assembled on an area and geographical basis.

The results of each completed audit-programme, though relative to external risk, could be organised according to a well known method of representing and comparing risk—for certification purposes, refined and adapted. Thus, we would form the *risk matrix*. This would be a support tool for decisions and it aims to act as comparison of the various features of different risks, contexts or situations. It works by approximating and subdividing through a simple discretization of Cartesian-plane into "iso-risk" surface areas, usually characterised by largely equivalent values for hazard (normally in ordinates) and magnitude or damage caused (usually in abscissae). For CEPRED, we will use this tool but adapting it to our work, i.e. we will not take into account the *magnitude* (or damage caused or damage potential) which is the combination of vulnerability and exposed value (or worth exposed) but the vulnerability alone—a choice deriving from our basic decision for the certification to use *specific risk* R_s.

Thus, we will a *specific risk* rating matrix. And on this we will mark two thresholds, known to risk analysts as *target threshold* and *limit threshold*. As applied in the pertinent literature, they represent the thresholds of risk *acceptability* and (transitory) *tolerability* (where the risk has to be reduced, usually, on the basis of the approach called ALARP—As Low As Reasonably Practicable) (cf. Lenoble-Durand 2011). In the CEPRED context, they split the specific risk matrix into three sections. In the first, the risk is *acceptable*, carrying 1 to 5 blue shields. In the second, the risk is not acceptable (grey-shield zone), the situation being deficient but capable of rising to acceptable values through essential intervention. The third section corresponds to *unacceptable* risk (red-striped shield) and only very radical moves (sometimes impossible) could improve things. The synthesis of audit-programme results in the risk-specific matrix can prove useful for a local or province civil-protection service, who, with the information clearly laid out in one unified scheme, can grasp the overall distribution of hazardousness and vulnerability for certified buildings. The results would be even more useful if the CEPRED analyses were systematically organised and carried out in collaboration with public bodies and public or private owners (also cooperating among themselves) grouped into neighbouring areas, such as town, blocks, town districts or commune areas.

As regards the Lecco analyses, the synthesis resulting from the specific-risk matrix shown in Fig. 3.14, one sees that there is a further possibility of *interpreting* the condition of territories. And this derives from the small sample examined; telling us that in Lecco Province the local hazard is more of a problem than the buildings' vulnerability.

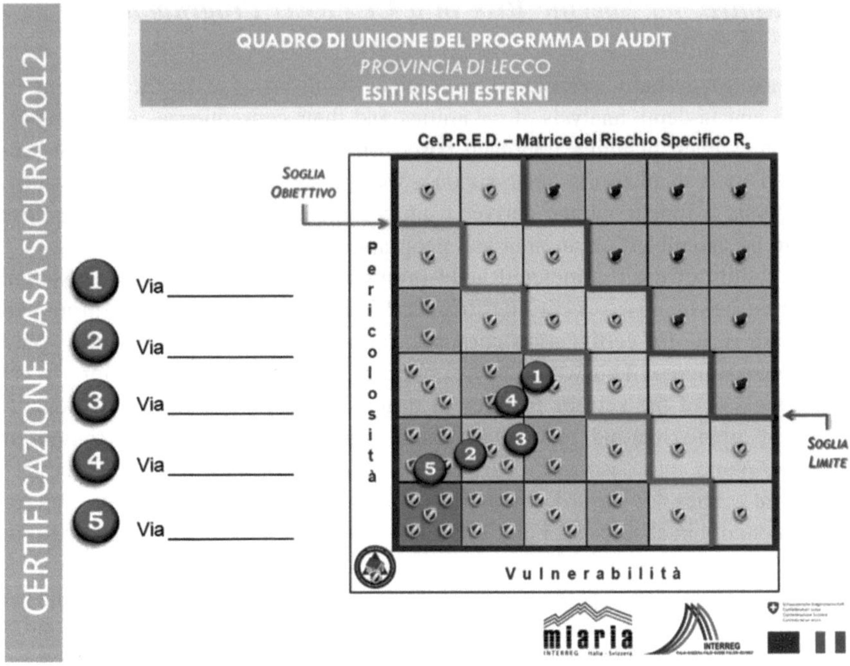

Fig. 3.14 The proposed unified summary of a PRED audit programme (for use in the Lecco Province) for the possible use of the local civil-protection services

3.6 Composition of and Requirements for the Audit-Management Team

The auditor's role is that of a professional with qualifications and ability to complete an audit on the lines foreseen for certification. Thanks to these competences, on the basis of what is stated in the ISO 19011:2011 and or ISO/IEC 17021 and ISO 31000 norms, different auditors team, each working independently, should, if examining the same circumstances, produce the same results. So in advance, the auditor's efficiency and professionalism must be scrupulously established and confirmed—in addition, of course, to the auditing procedures (planning, execution, enquiries, etc.).

As described in the ISO/IEC 17021, competence criteria should be determined with regard to the requirements of each type of management system standard or specification, for each technical area, and for each function in the certification process. For instance, for an environmental management system standard, the term "technical area" is related to the categories of activities, products and services related to the environmental aspects affecting air, water, land, natural resources, flora, fauna and humans.

So that, the *minimum* qualities needed to carry out a "campaign" for a CEPRED audit programme correspond to some articulated features—and, at the moment, we can identify them as follows:

- A *senior* figure able to supervise a whole audit programme and the valuation for each building with its conformation. This person must be able to oversee, and control as a whole the disaster-risk methods and the problems of safety for geographic areas and buildings in which the PRED certification operates;
- A person experienced in valuation who could function as *leader* in organising and operating the audit programme. He should have background in technical-scientific disciplines concerned with natural or man-made or constructed environments and territory (engineering, architecture, geology, environment sciences, etc.);
- one or more experts in geographical information systems who can optimise, archive and update area risk-analyses at programme level and also insert such information into a second database of geographical data regarding valuations of single buildings (in the mentioned computer based tool PREDICT);
- One or more valuing teams, made up of, at least, two evaluators, and consisting of engineers or architecture graduates with qualifications in environmental, civil, territorial and building matters and having at least 5 years experience—they should preferably have complementary specialisations within the team, such as one being expert in structural and building disciplines and another in risk and territory.

In particular, the valuing teams will have attended specific training sessions to grasp all the certification-mechanisms and will have been through practical field training by following some real building-checks and the examined structures' complexity and context problems.

A sample of discipline-specific knowledge and skills of auditors in building disaster risk management should be sufficient to enable the auditor to examine the building and the disaster risk local framework and generate appropriate audit findings and conclusions. A selection of main discipline-specific knowledge and skills from the PRED point of view are:

- disaster risk terminology;
- disaster risk metrics and statistics;
- measurement science and monitoring techniques;
- interaction between buildings and environment;
- phenomenology of the disasters included in the PRED (e.g. seismic, hydraulic, meteorological, etc.);
- techniques for determining risk (e.g. environmental aspects/impacts evaluation, including methods for evaluating significance);
- quantitative risk assessment and risk analysis techniques;
- emergency management;
- building resistance performance evaluation;
- prevention, protection and mitigation techniques (at different spatial and temporary scales);
- identification, monitoring and forecasting risks and events.

3.7 Certification Feasibility, Running, Reliability and Continuing Improvement

The ISO/IEC 17021, as, for instance, the EU 2002/91/CE directive on energy performance in the building industry, prescribes that the body issuing the relevant certification should be independent and without any conflict of interest. We have seen that this third-party position is essential for CEPRED certification, too. This is all the more obvious when we recall that on the award of the positive blue shields CEPRED plaque the building's market value stays firm or rises.

The norms ISO 19011 and ISO/IEC 17021 foresee that an audit programme must also include "all the activities necessary for planning and organising audit types and numbers as well as providing the resources for the audits to be carried out efficiently within the fixed time limits". The norms also foresee a *high level* of organisation which ensures authority for overseeing the audit programme whose directors should "establish, put into practice, control, re-examine and improve the audit programme"; they should also "identify the necessary resources and ensure that these are made available". The flow or sequence of such audit-programme control is representable with the application of the *P-D-C-A*[10] *(Plan*[11]*-Do-Check-Act)* method, also known as the *Deming*[12] or *Control Circle*. With this method, the audit-programme controller defines the audit programme, applies it, maintains it, re-examines it and if necessary updates or improves it.

3.7.1 Resource, Growth and Spread of the Certification —Towards the CEPRED Agency

The procedures involved in activating certification in various areas are complex from the very start with the *initial definition* of the territory's hazard. This means,

[10] The P-D-C-A method is also the basis for procedures of environment control (ISO 14001-2004). To synthesise: "*Plan*" stands for planning aims and the action needed to achieve them; "*Do*" is putting these decisions into practice; "*Check*", continuing control of interventions, objectives reached and laws respected; "*Act*", confirmation of procedures if they have been successful or, if not, their correction or improvement—the latter should in any case be continual (Foglio 2008).

[11] Planning, in this case, does not refer to one of the n audit plans making up an audit programme; instead, it concerns the entire planning contents of the total programme.

[12] Edward Deming, an expert on procedure improvement methods, influenced the results obtained in Japan with his Toyota production system. He maintained that procedure improvement must be continual. Basically he was convinced that those responsible for much of a drop in quality and production were managers rather than the workers. (Ishikawa. K., (Lu. D. J. trans.), 1985, What is Total Quality Control? Prentice-Hall Inc., Englewood Cliffs, NJ).

of course, depending on the high professional qualities[13] of whoever is involved—which brings us back to the requirements of the 19011 and 17021 norms regarding the management "authority" that should guarantee the audit programmes' workability trustworthiness and completion.

We think the best way to reach our objectives is to turn to the system developed for energy certification. Two instances of this, out of any possible examples, are to be found in the Agenzia CasaClima (House Climate Agency) and the management service of the Lombard Region CENED provided by Cestec Spa (Cestec PLC).

According to this model, the CEPRED certification system should be run exclusively by a body which we will call *CEPRED Agency*. The publication you are reading at this moment is not interested in whether this agency is a private company or government body, public, private or mixed in ownership, or however set up. As well as taking on the job of encouraging the spread of certification, the Agency would guarantee support and know-how for a network of external certifiers listed, after selection, in a table produced by the Agency itself, which would keep an eye on their performance and behaviour and on the eventual fees paid.

The statements and plaques foreseen in the certification will be emitted only by the agency itself, as all the policy of governing certification mark (above all the authorizations to use it for certified clients). This will ensure control of the *unique progressive number* given to each building, the number that acts as a protection against eventual counterfeiting. This safeguard is tighter than that offered, for example, by some energy-certifications offices which make only sample checks.

The agency will also be able to form a brains trust of scientists and technicians to ensure the continual updating and improvement of the certification contents. Thus, it could widen its applications to structure-types that become evermore complex and varied. Such achievement may result, above all, in internationalizing the rating system devised and so wider and more open cooperation between Italy and the EU's or world scientific and administrative communities would be possible and achievable. With such development the Agency (or an international federation or network supporting national agencies) could be able to offer specific certification training for professionals such as designers, planners, builders, dealers in materials, etc. It would also organise an information service for owners, administrators and users and for public officials from local and regional authorities who are responsible for managing territories, social defence, building or civil protection schemes. At the same time, the agency organisation will inevitably have to provide software and data systems for managing certification systems, for storing all the collected information and for auditing.

[13] Think, for example, of the extreme problems such as the impossibility of standardising the sources of data to be used which depend on what is available territory-wide and which are notoriously inhomogeneous across the whole of Italy, not to speak of the whole of Europe.

Chapter 4
Auditing and Analytical Procedure for Risk Indices Assessment

Auditing and the analytical procedure for evaluating the risks weights, hazards, vulnerabilities and specific risks are carried out by developing the auditing phase described in the previous chapter and achieving the technical *outputs* and *indices* listed in Fig. 4.1.

4.1 Valuation Area-Hazard: Initial Survey

In applying an audit programme, CEPRED first of all sets up a survey of territorial hazard. The survey runs through three phases with large-scale coverage of the area threats of external risk.

As has already been shown, this activity is fundamentally important in achieving a close definition of an area's possible hazard framework. Thus, it is also indispensable and basic for formulating the *Index of edifice's resistance predisposition to external risks* $I_{res(ext)}$. As regards the expression (2.3) of the above index, the outputs for activities in the *initial survey of territorial hazard* are the p_i weightings. This means we can proceed from analysis of the various single specific risks to one figure as total value allowing us to compare different buildings in the same area while keeping in mind calamities-intensity. The formula for getting p_i weighting for "i-th" risk (or hazard) is (2.4), which, for clarity's sake, we repeat here:

$$p_i = \frac{\sum_{j=1}^{n} S_{ij} \times Cm_{ij}}{S_{tot}} \times C_{HS(i)}.$$

Phase 1, for auditing procedure, regards area definition and the audit programmes reference period. This means also fixing the final moment of t_f for data collection and obtaining the first useful output, the valuation of the total reference surface-area S_{tot}, normally expressed in km^2. In the three test cases that were made, the S_{tot} value coincided with the value for the entire province-area. But, as we have already pointed out, other choices could bring it to coincide with different

D. F. Bignami, *Towards a Territorial Multi-Disaster Buildings' Resistance Certification*, SpringerBriefs in Environmental Science, DOI: 10.1007/978-88-470-5223-9_4,

	Phases of PRED audit programmes	Quantitative Outputs	
AUDIT PROGRAMME (1 per reference area)	***1. PRED opening*** and definition of territorial *reference-area* and fixed *reference-period* (at programme scale);	S_{tot}, t_f	INITIAL FRAMEWORK OF TERRITORIAL HAZARD
	2. *Collection* of geographical, disaster hazards and past events data and documentation;	n, A_f	
	3. Initial (*desk*) analyses and evaluation of the extent of territorial hazards;	S_{ij}, Cm_{ij}, $C_{HS(i)}$, p_i	
	4. Preliminary-stage 1 (desk) analyses of the hazard site (local scale);	H_{iL} *temporary*	AUDIT STANDARD PLAN (1 per building)
	5. Preliminary-stage 1 (documental) analyses of the hazard and vulnerability of each building;	H_{iL}, V_G *temporary*	
	6. Direct-stage 2 (field) check on site and its locality's *hazard;*	H_{iL}, V_G, V_C *temporary*	
	7. Direct-stage 2 (visual) check on the building's *vulnerability* and on its *internal safety;*	H_{iL} V_{iL} V_G, V_C $R_{int(k)}$ *temporary*	
	8. (*desk*) *elaboration* of the data collected;	H_{iL}, V_{iL}, V_G, V_C, $R_{int(k)}$ *final* – $I_{RES(EXT)}$, $I_{RES(INT)}$, I_{RES}	
	9. PRED results published (with eventual suggesting for improving) and eventual assignment of the *Blue-shields plaque;*	***Blue-shields***	
	10. eventual ***PRED re-start*** of another audit for *re-checking* of resistance performances (re-certification).	S_{tot}, t_f	▲

Fig. 4.1 Phases of a CEPRED audit and its technical outputs

contexts, areas within a province, between different provinces or even regions—or nation-sized areas.

Phase 2 regards territory documentation of disaster risks in the programme-reference area. This means first of all fixing the number n of hazards i present in the area and, as reference, keeping the t_f moment which gives us the latest update of these data. Beyond this, further documentation occurring will inevitably only be used for later audit-re-examinations starting with a revision (phase 10) of the *initial survey*.

The documentation and the collected data will enable us to get a value for the index of *source reliability* A_f expressed as a percentage (of the ideal total

Hazard	Main data and basic sources	Availability / completeness / updating (Lecco prov.)
Seismic risk[a]	National and/or regional seismic zoning – municipal seismic micro-zoning	75%
Hydraulic risk	PAI zones and measurement of threatened areas; AVI catalogue – data on Dams, data from the analysis of the EU flood directive	75%
Chemical-industrial risk	At risk sites localised per town – PEE sites	75%
Transport risk	Suburban road network, rail network – Airport runways and flightpaths – Past hazardous goods accidents	75%
Forest-fire risk	Soil uses data – Past fires data	100%
Meteorological risk	Windforce data – temperature and air pressure – hail data – lightening data – snow – ground temperature – storm frequency (province and municipal scale) – altitude	50%
Hydrogeological risk, avalanches	Maps for probable localisation avalanches – soil uses – slope and height data	50%
Hidrogeological risk, landslides	IFFI data – soil uses	100%

Fig. 4.2 Scheme for valuing availability/completeness of information needed for index of *reliability of A_f sources*—sources for separate risks. [a]As the seismic hazard is relatively low, the limited territorial wide-ranging of micro-zonation in this case has been only lightly weighted

documentation) established with the available data and factors serving for certification. Figures 4.2 and 4.3 detail information deriving from material acquired for Lecco province in 2009. These data allowed us to arrive at the A_f index value, at the moment by an arithmetic mean calculation with respect to the best data one can imagine it available for a territory on the basis of rules in force at the time t_f. The same data could be also used to enrich the information in the pages concerning single territorial risk in the PRED statement, as already done in relation to main synthetic buildings predisposition of resistance indices.

Phase 3 of the initial survey of territorial hazard dedicates analysis and valuation to the extent of the territory's hazard—always contained within the programme's area-scale and up to the t_f reference moment.

Using the information collected in phase 2, by means of a GIS—Geographical Information System, reference territory is divided into zones of intensity-equivalent j magnitude for each of the n risk i considered, after the discretisation of the pertinent reference scales. Thus, all the different S_{ij} values are obtained for the area under scrutiny where an i-th risk of j-th intensity-class is present. There is then the fixing of various Cm_{ij} (coefficients of magnitude describing the i-th risk with the

Multi-risk sources	Administrative level / Data and basic sources	Availability / completeness / updating (Lecco prov.)
Foresight and Prevention Programme	Region – Province	75%
Emergency Plan	Province – Municipality	80%
SPATIAL PLANS	PROVINCE	70%
Land use Plans	MUNICIPALITY (such as territorial government plan)	75%

Fig. 4.3 Scheme for valuing availability/completeness of information needed for index of *reliability of* A_f *sources*—multi-risk sources

j-th intensity-class) and the values for *past-incidence coefficients* $C_{HS(i)}$. This move brings out the various p_i weighting parameters for each of the n risks present in the programme's reference territory.

4.2 The Programme Scale and External-Risk Weighting

As already noted, in principle, though with a few variations for some risks (see later), the hazardousness classes set coefficient-magnitude values Cm_{ij} which can vary *linearly* from 1 to 5 according to their intensity (since, if present, a risk cannot be eliminated). The *range* and distribution of these values are valid only *provisionally*—seeing that the proposed certification is still experimental. These values were fixed through a *sensitivity analysis* invented for the purpose of the calibration of our rating and labelling system. But they are still being studied and checked, in Italy, too, since we need data-sets that are fuller and geographically wider-based if they are to be valid (this process will be probably necessary in each new country in which PRED certification will be activated). The p_i data sets obtained come from a *subjective*[1] method which is, however, rigorous in calculation and in repeatable procedure. These sets could be different. On the one hand, this is possible in cases with a wide range, for example, from 1 to 10, which would lead to heavier weighting of the more intense classes, while the coefficients for the minor classes would vary, in the lowest case being always equal to 1 or would show very little variation in the neighbouring classes of lesser intensity. On the other hand, these could also be different in cases of non-linear distribution—for instance, exponential or logarithmic for values within the range, with some coherence to physical scales of magnitude for certain risks.

[1] As explained in Chap. 2, equivalent weightings would have compromised certification-efficiency. *Unjustified weightings* made by experts would not have allowed a homogeneous coverage of the various areas.

With the aims of our research in mind, however, these possibilities, though interesting, have still not been exploited. There are two reasons for this. One is the limited range of the intensity scale for the calamities considered with data now available, because they are referred to relatively safe territory (for instance regarding seismicity). The other reason is the results obtained which, overall, were not noticeably different from these for the various simulations set up for sensitivity analysis. However, next investigation and calibration tests will involve more "difficult territories" and verify different possible progressions of p_i values, to be confronted with the *linear solution* "1–5" until up to now provisionally adopted—different possible hypotheses are *linear* "1–10" or *exponential* "1–10" (the latter practically being an intermediate progression between two linear ones). This verification could also include the possibility to adopt variable solutions and choices of the progression of p_i values, which are to be preset depending on the related i-th hazard (Fig. 4.4).

Obviously, following professional practice and so as to be comparable, the calculation of p_i weighting obtained like this requires a final normalisation later. For any risk foreseen in certification until up to now, we will describe the procedures adopted for getting each of the n weighting parameters p_i, on the basis of which we has obtained the three following lists regarding the Milano, Lecco and Verbano-Cusio-Ossola provinces.

Data in Fig. 4.5 show a consistency with respect to what one could predict about the three provinces. The metropolitan area of Milan is an urbanised territory, with scarcity of forests, heavy traffic, industrial productions and low seismicity and all these factors contribute to characterise its p_i parameters' list. Lecco and Verbano-Cusio-Ossola are mountains territories, with many forests, heavy slopes, low seismicity, fewer inhabitants and industries than Milan.

4.2.1 Seismic Risk

For seismic risk, the basic division into zones was done according to the values of the Italian seismic hazard map for peak horizontal acceleration, with 10 % probability of non-exceedance in 50 years, referred to average hard ground conditions, (Ordinance of the President of the Council of Ministers 20th March 2003 "First steps in general criteria for national-territory seismic classification and for normative techniques for seismic-zone constructions"). This basic criterion was refined later, where possible, thanks to the availability of better-detailed information. This was true of the Lombardy Region, which, in its PRIM—Programma integrato di Mitigazione dei Rischi Maggiori 2007 made a division of zones 1, 2 and 3 into four subzones. Figure 4.6 shows the assignment of values for magnitude coefficient. Note that on the basis of the valuation of the available data for horizontal acceleration when the subdivision was into only 4 classes, each class was given the greater value of the subzones, except the fourth.

For seismic risk hazard, the *historical-incidence coefficient* $C_{HS(seismic)}$ for Italy was fixed as equal to 1 as the historical-incidence value since seismic zoning is

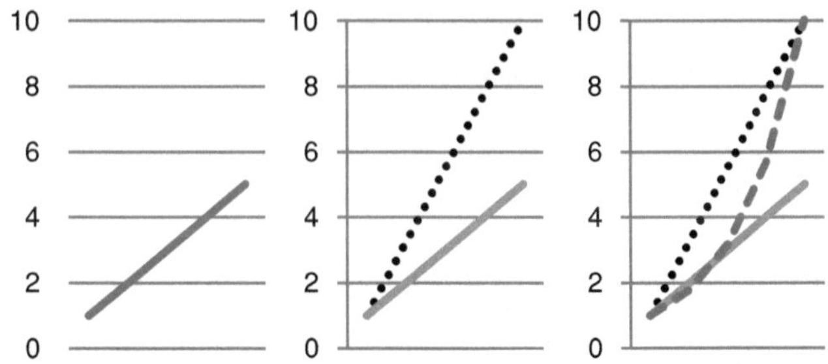

Fig. 4.4 PRED's possible p_i values progressions: *linear* "1–5" (*left*—until up to now provisionally adopted), *linear* "1–10" (*centre*) and *exponential* "1–10" (*right*)

Milano			Lecco			Verbano-Cusio-Ossola		
HAZARDS	p_i	$p_{i\%}$	*HAZARDS*	p_i	$p_{i\%}$	*HAZARDS*	p_i	$p_{i\%}$
1.meteorological	1,79	25	1.landslides	1,98	26	*1.hydraulic*	5,5	35
2.transports	1,44	20	*2.hydraulic*	1,34	18	2.landslides	2,8	18
3.hydraulic	1,41	19	*3.meteorological*	1,21	16	*3.forest-fires*	1,53	9,9
4.industrial	1,27	18	*4.forest-fires*	0,72	9,4	*4.meteorological*	1,51	9,7
5.seismic	1,16	16	*5.transports*	0,72	9,4	*5.transports*	1,45	9,3
6.forest-fires	0,16	2	*6.industrial*	0,67	8,8	*6.avalanches*	1,41	9,0
-			*7.avalanches*	0,56	7,3	*7.seismic*	1,15	7,4
-			*8.seismic*	0,45	5,9	*8.industrial*	0,20	1,3

Fig. 4.5 PRED lists of territorial hazard regarding Milano (*left*), Lecco (*centre*) and Verbano-Cusio-Ossola (*right*) provinces

already based on historical, i.e. past criteria (cf. Petrini 2009). As a result, in our case (2.4) becomes:

$$p_{i(seismic)} = \frac{\sum_{j=1}^{n} S_{ij} \times Cm_{ij}}{S_{tot}}. \quad (4.1)$$

More generally, a similar formulation could be valid in many other countries, since seismic zones are defined on the basis of the same principles as in Italy (above all for most of the applications of EN 1998 cf. Solomos et al. 2008). For instance, in France, in Portugal[2] and in Slovenia[3] we could try to start to calibrate our methodology with the following magnitude coefficients Cm_{ij}. These starting values

[2] Regulamento de Segurança e Acções para Estruturas de Edificios e Pontes, Decreto-Lei N.o 235/83, de 31 de Maio.

[3] Ministry of the Environment and Spatial Planning, Geophysical Survey of Slovenia, "Tolmač karte potresne nevarnosti Slovenije", Janez Lapajne, Barbara Šket Motnikar, Polona Zupančič, MOP—Agencija RS za okolje, Urad za seizmologijo, Dunajska 47/VII, 1000 Ljubljana, www.arso.gov.si/podro ~ cja/potresi/podatki/projektni_pospesek_tal.html.

Zone / ***subzone***	**Horizontal acceleration probably over 10% in 50 years [ag/g]**	***Cm$_{ij}$***
1		5
1d	*0,325 - 0,350*	*5*
1c	*0,300- 0,325*	*4,67*
1b	*0,275 - 0,300*	*4,33*
1a	*0,250 - 0,275*	*4,00*
2		3,67
2d	*0,225 - 0,250*	*3,67*
2c	*0,200 - 0,225*	*3,33*
2b	*0,175 - 0,200*	*3,00*
2a	*0,150 - 0,175*	*2,67*
3		2,33
3d	*0,125 - 0,150*	*2,33*
3c	*0,100 - 0,125*	*2,00*
3b	*0,075 - 0,100*	*1,66*
3a	*0,050 - 0,075*	*1,33*
4	0,050	1

Fig. 4.6 Values for magnitude coefficients for seismic risk

have been thought maintaining a linear proportionality among the related acceleration values (it is obviously possible, at the same time, to test different solution, as shown in Fig. 4.3). Logically, as there are still differences among national seismic provisions (despite the recommendations for harmonization at European level), consequently, it is, initially, for PRED certification, too, being an interesting field for future investigations and for the development of PRED itself (Fig. 4.7).

For the Verbano-Cusio-Ossola Province, we tried out a *variation* regarding the surface area for (4.1). This was done because of the vast part of the province's territory that was not built-up, that is not urbanised. This fact had to be taken into account—otherwise the resulting data would have been distorted. This situation is caused by the forming areas (or sparsely-inhabited pasture-land) which are hilly or mountainous. Another factor here, though it is not a great problem, is the province's non-uniform spread of seismic hazard. Thanks to the use of GIS, it was decided to take no more than the built-up areas into consideration, in accordance with the later simple, intuitive transformation which replaced the overall surfaces with the urbanised ones $S_{(urb)}$:

$$p^*_{i(seismic)} = \frac{\sum_{j=1}^{n} S_{(urb)ij} \times Cm_{ij}}{S_{(urb)tot}}. \quad (4.2)$$

Zone /	***subzone***	**Acceleration values**				Cm_{ij} *(linear hypothesis)*			
		Italy-2003	France-2000	Portugal–1983-action type 1	Slovenia-2002	Italy	France *(starting values to be calibrated)*	Portugal	Slovenia
1/III/A/red		0,35	0,36	0,48	0,25	5	5	5	5
2/II/B	or.	0,25	0,25	0,34	0,225	3,67	3,31	3,35	4,33
	l.o.				0,200				3,67
	ye.				0,175				3
3/Ib/C	l.g.	0,15	0,15	0,24	0,15	2,33	2,77	2,18	2,33
	gr.				0,125				1,67
4/Ia/D/d.g.		0,05	0,10	0,14	0,1	1	1	1	1

Fig. 4.7 Values for PRED magnitude coefficients for seismic risk—comparison with Italy and hypotheses of starting values (to be calibrated) for France, Portugal and Slovenia

This modification was also interesting and promising for all the areas with a low overall population-density (about which we can imagine to fixe an overall threshold) but, at the same time, with a highly concentrated population in a few centres. Such areas could also have an asymmetric risk-distribution—and not only in the seismic component of the certification proposal. In some cases, one might exclude only the mountainous territory while including pastures and planted fields. In a different manner, in handling these cases, one could define more precisely the initial reference territory (S_{tot}).

When the results of micro seismic zoning[4] have become more available and widely-known (which, in Italy, nowadays, still are not available), these valuations will definitely increase and will serve in better valuation of the inner-province scale.

4.2.2 *Hydraulic Risk*

For hydraulic risk, the territory's zone-division into areas with similar magnitude of threat took, as basic reference in Italy, the information given in the Hydrographic Basin Planning document in as far as it was concerned with defining river-fluvial or flood-prone areas (A, B and C) and areas with different likelihoods of flooding (high, moderate or low).[5] In EU countries, the PRED procedure could use

[4] Explicit reference is made to this in the new technical norms for building. They specify the need to value the local seismic possibility as regards the landsite situation.

[5] In reference to legislation on Soil Defence (from Law 183/89 onwards) and to the "Act of coordination and management" (DPCM, 29th September 1998), which regard the urgently-passed legal section of the *Plan* for the hydrogeological situation defined in identifying and measuring areas at hydrogeological risk.

the results of the EU flood directive 2007/60/CE adoption, if they will prove to be homogeneous enough. This would have required to Member States to map the flood extent and to draw up related maps by 2013.

This choice involves care in fixing reference-territory for the audit-programme—particularly checking that this area does not cover possible minor portions under the management of different Basin Authorities who may have applied heterogeneous criteria in defining possible flood maps.

Moreover, sadly, it is common knowledge that job of defining the river-sectors has not yet be completed homogeneously across the whole of Italy and the problem is thought to be potentially increased by extending the application of the rating system adopted and proposed by us. For the most part, it has only covered the main rivers. This is why it was decided to have, for hydraulic hazard, a value different from 1 of the historical-social incidence-coefficient $C_{HS(hydraulic)}$ when the available data for maps were very incomplete—as is true in the vast majority of cases. For calculating the $C_{HS(hydraulic)}$, we propose to use the historical data on flooding in the AVI catalogue (these data refer to the period 1810-2000).[6] As a result, in the specific case of hydraulic risk, the (2.4) formula is:

$$p_{i(hydraulic)} = \frac{\left(\sum_{j=1}^{n} S_{ij} \times Cm_{ij}\right)}{S_{tot}} \times [1 + (\sum_{k=1}^{m} S_{(fl)k})]/S_{tot}, \qquad (4.3)$$

In this formula, we take into account the surface k flooded$S_{(fl)k}$ in each commune m, during the time considered in order to get a coefficient for historical incidence $C_{HS(hydraulic)}$, which, here, as it is formulated, represents a factor of increasing and combining all the events that characterise the reference-territory. Note how the use of a multiplying coefficient where k (representing data of records in the AVI catalogue) is coherent with the chance of overlapping among historically-flooded areas and parts of the river areas included in the first term, coherent, too, with the chance that some areas have been flooded several times in the time considered. Figure 4.8 gives the coefficient values Cm_{ij} for the various sectors or zones floodable from river or lake.

If the necessary data are available, the C sector and zones with high flooding-probability can be divided into two classes, thus making a class of value *4*. In the sector with greatest intensity, we would leave the situations with *more water-depth, greater flow velocity and more sediment transportation*. The other sectors can be left in the newly-formed lower classes. Similar divisions based on the above-mentioned parameters could be made for the remaining classes—but as always depending on acceptability, accurate data being available for the entire programme's reference-territory.

---

[6] Project for Damaged Italian Areas (AVI-Aree Vulnerate Italiane) archiving information collected by the CNR National Group for Defence from Hydrogeological Catastrophes.

Hydraulic risk	**Cm_{ij}**
"C" areas or zones with high probability of flooding	5
"C"areas or zones with high flooding-probability because of minor water-depths, flow velocity and amount of sediment transportation (if data are available)	*5* *(4)*
"B"areas or zones with moderate flooding-probability	3
Flooding from lakes in subjected areas	2
"A"areas with low flooding probability	1

Fig. 4.8 Values for magnitude coefficients for hydraulic risk

Turning to the problem of dams or artificial barriers in the reference territory, even though related flood hazard maps, in case of break, are not easy to collect (even in Italy, in spite of the DSTN/2/7019–1996 norm on the dam emergency planning) one can only take into account mapping produced with the specifically devised software so as to form a calculation of profiles of water-surface which are, then, inserted as an input into the GIS software managing Digital Elevation Model–DEM and dealt with in the same way as the natural floodable areas, figuring together with the latter with the same classification.

4.2.3 Chemical-Industrial Risk

Man-made disasters (and "NaTech—Natural hazard triggering Technological accident" disasters) can occur more often, above all due to the increased complexity and interdependency of industrial systems and to human activities. In part, they are also due to the increased frequency of some kinds of extreme natural events (for instance, meteorological, hydrological or forest fires). Among several kinds of man-made disasters, having in mind the edifices as a target, the chemical-industrial risk, along with the risk of transportation of dangerous goods as well as proximity to transportation routes (see next paragraph) are the most relevant of all (taking into consideration that dam-break is included in the hydrological risk and nuclear risks are not included in our rating system).

For chemical-industrial risk hazard resulting from the use or storing of hazardous substances, the basic reference-source for dividing territory into dangerous classes is the zoning maps of factories at major-accident hazard (in Italy from articles 6 and 8 of Law 334/1999 in response to the European directive "Seveso II" 86/92/CE—lower and upper tier establishment).

For each of these hazard areas, every prefecture sets up an emergency plan called EEP—External Emergency Plan. These plans usually define three zones around each installation; they are known as zone 1, "*lethal effects*", zone 2 "*serious injuries*" and zone 3 "*reversible effects*". Furthermore, in France, for instance, on the basis of the already mentioned PPRT, more advanced regulated zones have also been established, being related to the problem of the urbanisation in risk areas. They are known either as *ban zones*, or red zones (*R* and *r*, to be expropriated or relinquished), or as *limitation zones*, or blue zones (*B* and *b*, external to ban zones). This is particularly interesting since these protective measures on the buildings (future or existing ones) could become compulsory to reduce the frequency occurrence of consequences of accidents (i.e. decreasing vulnerability).

However, it is clear from some past industrial accidents, including recent ones like Toulouse 2001, that this zone definition is, unfortunately, influenced by a number of scenarios (applied by involved professionals) which are too limited when compared with what may really occur and, at times, too varied in method for dealing with the whole territory of a province—considering the choices and decisions made for each separate site. And sadly, access to this kind of information is not as easy as it should be. Despite the recently adopted Seveso III Directive 2012/18/EU obliges operators to include in their *safety reports* (to be made available to the public upon request) a description of any technical and non-technical measures relevant for the reduction of the impact of a major accident.

So, for our certification, we had to turn to the most homogeneous information available for the whole province's/district's reference territory. In particular, as regards Milan,[7] it was only possible to make an aggregate accounting for no more than municipal areas (and not of the different External Emergency Plan zones), though, apparently, with the latest and complete data. The relevant *External Emergency Plan* gives a hazardousness-classification for communes, based on the presence or nearness of factories subject to art. 8[8] (total notification), art. 6 (simple notification), and art. 5 paragraph 3 (simple report) of the Seveso II directive.

In any case, this division was accepted as an adequate information-base for a starting point in finding magnitude-coefficient values. The 5 classes of hazard were thus defined (as shown in Fig. 4.9)—trying to apply criteria that also covered the uncertainties of the types of scenario considered and their reference-intensities.

A similar type of classification was set up for Verbano-Cusio-Ossola. But in this case we tried out a classification based on the detailed information available in the Piedmont Region's "Information System on Activities with major accident hazards"—SIAR, *Sistema Informativo delle Attività a Rischio di incidente rilevante.* Actually, the SIAR information means we can, in aggregate, define the overall areas for every type of External Emergency Plan zone (realised) and the *frequencies* (hypothetical) of possible industrial accident phenomena. By combining these two sources in a double-entry matrix, we were able to produce a classification analogous to the Milan-Province one—but on a more refined basis, taking into account both the seriousness of effects and their possible frequency—just as is done, in effect, to obtain a risk matrix.

The SIAR data also provide positioning for *hazard centres*, on whose basis one can reformulate the classification also on sub-municipal area basis. This means we can also take account of the communes neighbouring, those with industrial plant at

---

[7] At the time of this research, Milan still included the latest Monza-Brianza Province.

[8] These factories are obliged to submit a *safety report*—which should contain all the information needed for our valuation.

Chemical-industrial risk	Cm_{ij}
Municipal areas with maximum hazardousness *(or lethal effects zones or French R zones)*	5
Municipal areas with high hazardousness *(or serious injuries zones or French r zones)*	4
Municipal areas with medium-high hazardousness *(or reversible effects zonesor French B zones)*	3
Municipal areas with medium hazardousness *(or municipal zones outside the External Emergency Plan or French b zones)*	2
Municipal areas with low hazardousness and/or neighbouring outskirts communes[a]	1

Fig. 4.9 Magnitude-coefficient values for chemical-industrial risk. [a]In its first application (Bignami, 31st March 2009, Valuation of risk and safety of public and private buildings, Milan Polytechnic, Auditorium), the outskirts communes were given an 0.5 value, which in later-value-revisions was judged too cautious

risk and of the commune areas outside the External-Emergency-Plan. As for the p_i weighting parameter calculation for Verbano-Cusio-Ossola, we made an effort to account for both the factors—the commune zones involved and the communes neighbouring on those with at risk industrial plant, i.e. combining the commune directly involved and their neighbours. But the values thus obtained, which were potentially more precise, proved negligible and insignificant effect on overall multi-criteria valuation—given that the province's specific risk is hardly relevant. What is more, doubts still linger about the non-homogeneous criteria used in defining the risk zones. So the division for Milan remains confirmed as a reference, though still improvable, since it is more readily homogeneous and valid in any territory-context. Moreover, it has the likely merit of enabling us to over-calculate rather than under-calculate. As for seismic risks and hydraulic ones, it is worth noting that in the above case, too, the availability of more detailed information and results from deeper research usually allow the values reached to be smaller.

For chemical-industrial risk hazard, the *historical-social incidence coefficient* was taken to be equal to a single unit ($C_{HS(chemical\text{-}industrial)} = 1$) since the very varied types of risk involved, the plants' continual technical and technological advances and the substitution or limited working-life of the plants themselves exclude any reference to or dependence on historical, i.e. past events. Therefore, in our case, (2.4) becomes:

$$p_{i(chemical-industrial)} = \frac{\left(\sum_{j=1}^{n} S_{ij} \times Cm_{ij}\right) + \left(\sum_{k=1}^{l} S_k\right)}{S_{tot}}, \qquad (4.4)$$

where S_k is the k-th areas of l neighbouring communes and S_{ij} is the areas of n communes (or n zones of the External Emergency Plan) and the municipal areas outside the EEP zones. These are valued on the basis of the defined magnitude-intensity j scale.

4.2.4 Risk of Transportation of Dangerous Goods as Well as Proximity to Transportation Routes

This type of risk refers to direct possible consequences from moving vehicles or from the goods they are carrying. As regards the *risk of transportation of dangerous goods as well as proximity to transportation routes,* one must also of course include all the various types of communication—and transport network present in the programme's reference territory as well as the relevant data for the flow of traffic and goods. So we took into account transport by *road, rail, air, and conduit (or pipes*), with the exception of shipping, that, for the PRED, will be probably included in *sea-coast* risk.

For the component *direct* consequences of hazard for buildings' *near transport routes*, the first aspect tackled was *road transport*. This was because inside a town or city, we have speed limits, traffic volumes, parked vehicles and features of inclines, bends, road repairs, weather conditions and traffic signals, and taken together all these factors make any collision between vehicle and building highly unlikely or most extremely rare. With this in mind, we used for Italy—and in the same manner it might be adopted in other countries—the Istat (National Statistics Institute) data for the provinces' road network (and above). This source gave us data for the road-length as well as the number and width of motorway and state-road lanes for the Regions, Provinces and Municipalities.[9] With the available data, a simple calculation gave us the surface involved and we assigned a *double damage strip-zone* (one for each lane-side) measuring 50 m wide. This extension, together with the road width, is the dimension to be used to obtain the total area at risk as regards nearness to roads which will then be given magnitude coefficients.

For *rail transport*, the procedure adopted was like that for the roads, with the same size damage-belts.

For *air-transport*, valuation was made covering communes involved in main *take-off routes* or in minor routes. This was true for different types of airport—national, international and intercontinental, tourist, sporting and military (main or secondary). For Milan Province, the communes involved in these flights-paths were those concerned with take-off and landing at Linate, Malpensa and Bresso.[10] Bearing in mind the special character of this transport risk component and problem of accounting for damage belts along take-off paths, the magnitude coefficients represent an exception in our developed procedure and we fixed them at a value *below* the single unit (one order of magnitude smaller).

---

[9] On the basis of the Decree 19th April 2006, "Functional and geometric norms for constructing road intersections".

[10] For the PRED, at his moment, we have not included the increase in hazardousness brought by nearness to transport routes of *heliports*. Their traffic appears negligible, in view of the approximations made in our procedure, and as regards the final results of the multi-criteria analysis.

For the hazard component resulting from the *transport of hazardous goods*, which is doubtless bound up to some degree with the data for hazardousness near transport roads, we made two considerations: this first regarded the need to cover hazardous substances moving through *conduits* such as pipelines for methane, oil or gas. For this calculation, we again took into account the length in kilometres of the main distribution network[11] and a damage strip-zone analogous to the one for road and rail networks.

The second consideration concerned the *historical data for incidents* involving hazardous-goods transport in the reference territory. Such data can help in reconstructing a hazardousness valuation that also includes specific traffic-flows. To calculate the relevant historical incidence coefficient, we considered a circle with a 1 km radius as surface (3,142 km^2 extra for every accident) for every accident in the *ten years* before the period (t_f) fixed for programme reference, when there had been only one accident in a commune. If there had been two or more accidents in one commune, then the 1 km radius (r) was doubled or tripled (and so on) so as to give more weight for the repetition. This treatment, in all likelihood, marks a persistent peril rather than a chance incident and so is to be valued more than linearly. For Milan Province, this calculation, unfortunately, rose to a value of 911 km^2. In view of these valuations, the calculation of (2.4) becomes:

$$p_{i(transport)} = \frac{\left(\sum_{j=1}^{n} S_{ij} \times Cm_{ij}\right)}{S_{tot}} \times [1 + (\sum_{k=1}^{q} S_{HA(k)})]/S_{tot}), \qquad (4.5)$$

where $S_{HA(k)}$ stands for the corrective *surface for past accidents* for the k-th of the q communes where the accidents took place. This value is easily arrived at, taking n_k as the number of accidents in the commune k, we apply:

$$S_{IS(k)} = \pi(n_k)^2.$$

After obtaining all the sought for surfaces (which, for air traffic, are commune surfaces) and the past-accident coefficient, the magnitude coefficients Cm_{ij} were given as follows (Fig. 4.10).

4.2.5 Meteorological Risk

Meteorological risk, on the basis of our classification, covers various phenomena: *snow, hail, storms and thunderbolts,*[12] *ice and freezing conditions.* A foreseeable next step of investigation will be devoted to the study of *extreme temperatures'*

[11] The capillary gas-network for each building comes under the internal risk of *explosion*.

[12] Differently from (Bignami 2011), where storms and thunderbolts were separate.

Transport Risk	Cm_{ij}
Motorways (and high-flow roads)	5
State roads	4
Province-, region-roads	3
Commune, municipal roads	2
Railways	2
Pipelines – methane, oil, gas	1
Air traffic: main take-off flight-paths (national, international, intercontinental, main military, tourist, sporting or secondary military airports)	0,5
Air transport: secondary take-off paths (national, international, intercontinental and main military airports) and tourist, sporting or secondary military airports.	0,1

Fig. 4.10 Magnitude coefficient values for transport risk

problems, related to *heat wave*, trying to assess its relevance and position in relation to our certification system. One could think to evaluate/consider the extreme temperatures problem as a splint-off of the ice and freezing conditions' issues, as it happens in the reverse weather conditions.

As a type of disaster risk, the meteorological one could be considered anomalous, or not so relevant. On the contrary, the probability of most types of extreme weather events is expected to grow significantly. Furthermore, the effect of climate change, in the long-term, could have a relevant impact and the potential losses could be highly dependent on the weight of exposure and vulnerability[13] of edifices to meteorological risk.

To get a single result covering all the above phenomenological classes of meteorological risk hazards, we used a further multi-criteria aggregate to represent their co-presence in a given territory. In this case, each component was given an equivalent weighting parameter, thus obtaining an average value for the components.

This decision was taken considering the relative "destructive" capacity of each meteo-risk phenomenon on Italian soil. A different choice, that had kept the various phenomena apart would have brought an overestimate of the effects. However, we must admit that a choice like this would need a re-think, if we were dealing with countries different from Italy or from the *Euro-Mediterranean* area,

[13] See Intergovernmental Panel on Climate Change (2012), Chap. 3: *Changes in Climate Extremes and their Impacts on the Natural Physical Environment in Managing the Risks of Extreme Events and Disasters to Advance Climate Change Adaptation.*

such as those that suffer from *tornadoes or typhoons* or other types of meteo-risk (for instance *dust-* or *sand-storm*).

For this specific group of risk, since calculations always refer directly either to a past "historical" series of events or to atmospheric physics generating the phenomena, possible territory magnitudes, $C_{HS(meteo)}$ will always be taken as equal to a single unit. This means clearly that (2.4) becomes:

$$p_{i(meteo)} = \frac{\sum_{h=1}^{5} \sum_{j=1}^{n} S_{(h)ij} \times Cm_{(h)ij}}{S_{tot}}, \qquad (4.6)$$

where $S(h)_{ij}$ stands for the n surfaces S of intensity j, relative variably to each of the 5 sub-risks h of the meteo-risk (i). And $Cm(h)_{ij}$ represents the *relative* magnitude coefficients of the 5 sub-risks h of meteo-risk (i).

4.2.5.1 Wind

Hazardousness from wind effects does not include storm phenomena, which will be dealt with later, but does include tornado effects. This hazard comes from *wind-pressure*, usually exerted horizontally on a building's. It is linked to wind with speed higher than certain speed-thresholds and to the territory-distribution of anemometer data. The latter can generally be taken from localised points, of which the values are marked in the corresponding Thiessen polygon.[14]

Having obtained the available data, one can make a valuation based on an intensity scale classifying values according to speed. For this purpose, one can refer in the field literature, to Beaufort[15] and his well-known valuation. This concerns the reference speeds v_b for windforce on civil buildings, a mapping drawn up by the National Research Council.[16] Using both these document-references, after trials with bigger and smaller values, and after checks on the choices of the Italian Governmental Civil Protection Department (with thresholds from 11.3 to 24.5 m/s or from 40 to 89 km/h) and after checks on the Fujita damage scale for rating tornado (in which level 0, corresponding to the level 8 of Beaufort scale of 62 km/h, is the starting no-damage level and 117 km/h the first level of

[14] The application in meteorology by the American Alfred H Thiessen of the diagrams of *Georgi Fedoseevich Voronoi* (Russian mathematician). Having chosen points of interest on a plane, the diagrams divide the space into areas where all the area-points are closer to the chosen point of interest inside the diagrams than to any other external point of interest.

[15] Francis Beaufort, British Admiral (1774–1857) proposed his 13° wind-scale, from *calm* to *hurricane*, in 1806.

[16] For instance, the project named "Identification of areas with meteo- and flood-risk" carried out by the Italian Biometeorology Institute (the IBIMET of Florence) and by the Information System for Hail, Flood and Wind (SI_GAV).

WINDSPEED	*Theoretical absolute* Cm_{ij}	*Relative* $Cm_{(h)ij}$
< 11,3 m/s (< 40 km/h)	*1*	0,2
11,3–17,1 m/s (40-61 km/h, *strong breeze to moderate gale*)	*1,8*	0,36
17,2–22,5 m/s (62–81 km/h, *fresh to strong gale*)	*2,6*	0,52
22,6–27,5 m/s (82- 99 km/h, *strong gale to storm*)	*3,4*	0,68
27,6–32,5 m/s (100 - 117 km/h, *violent storm to hurricane*)	*4,2*	0,84
> 32,6 m/s (> 117 km/h, level F1 of the tornado rating)	*5*	1,00

Fig. 4.11 Magnitude coefficient values for sub-risk "wind"

tornados cf. Fujita 1971; www.spc.noaa.gov/efscale/—Storm Prediction Center, the new division shown in Fig. 4.11 has been decided on for CEPRED[17]:

For assigning value to the polygon, it was decided to be prudent and refer to the maximum value to be found in the records for, at least, the last 10 years (if data were available) or to the maximum value, in case of average data, increased of a corrective value equal to their deviation standard σ, obtained on the basis of the hypothesis of a *lognormal* density function (cf. Kottegoda-Rosso 2008). The idea of getting the Cm_{ij} value by weighting calculated from value frequency seems, as for other risks like the seismic one, over-cautious and so was refused.

Obviously, this is maybe the clearest case in which differences, in standard event intensity and in edifices resistance characteristics, could benefit from a deep recalibration of the above proposed values, considering the regional geographic localisation of the active PRED certification programme.

4.2.5.2 Snow

Hazard from snow is, above all, due to *vertical weight* on roofs. Italian technical norms on *typical values for snowfall on the ground* mark out zones for all Italy, applying weight values correlated to fixed quotas (as are in many other international cases). These are mainly in proportion to the quota square of the measure above the sea-level, fixed on a basis of a fifty-year return period.

The first threshold fixed by CEPRED, in accord with the norm definitions, is 200 ms asl, the latter being an intermediate level of the 1,500 ms asl fixed by norm as limit beyond which snow values may be so high and so variable that they require special valuation of the local conditions. 1,500 m was chosen as the third threshold. The relevant Cm_{ij} values were proposed for climate zones, foreseen in the technical building-norms, we have already described in our calibration campaign (I—Alpina; I—Mediterranea). For these values see Fig. 4.12:

[17] Until 2013 the classes were defined with little differences.

Altitude above sea level (asl)	Theoretical absolute Cm_{ij}	Relative $Cm_{(h)ij}$
< 200 m	*1*	0,2
200-750 m	*2,33*	0,47
750–1500 m	*3,67*	0,73
> 1500 m	*5*	1,00

Fig. 4.12 Magnitude coefficient values for sub-risk snow

More detailed considerations on climate zones in Italy (and Europe) will, if necessary, be made about future applications. These will refer to past series of snowfalls (to obtain snow-load values) so as to create territory divisions for polygons and not for curves in levels. The level of improvement in the model's approximation, for territories already dealt with, will be secondary but it will definitely be more notable and opportune for territories with different climates or for those within provinces or districts.

4.2.5.3 Hail

Hazard for *hail* is mainly due to the density of the hailstone mass, the speed of the hailfall and how often hailstorms occur. There is an extreme variability in these phenomena and, sadly, data for their past occurrence are rare and not always reliable.[18]

If such data are not available, the frequency of meteorological conditions leading to hailfall have to be calculated by making an estimate of the *Wet Bulb Zero Temperature index*, which assigns probability values to hail events according to altitude and the consideration of the rapport between solar radiation and relative humidity, temperature and air-pressure. This information, though approximate, can be had from the territory's weather stations, which, if necessary, can be given the respective Thiessen polygons.

There is some reliability in the frequency data for hailstone-diameter classifications in Italy. This is on a ten-year basis. Even though they are not always effective for spatial aggregations, if they are on a level for an area smaller than a province. But thanks to, for instance, the data provided by the Bio-Meteorological Institute of the National Research Council, we have some reliable figures. With these, the relative Cm_{ij} values were obtained as in Fig. 4.13:

[18] The instrument for noting and measuring hail is a "grêlomètre" (French), "grelimetro" (Italian) (= "hailometer") and is less well-known than other meteo instruments. It consists of a polyester panel, square or rectangular having 15–40 cms per side. It is covered in a very thin layer of aluminium and has to be set out horizontally for the elements at about a metre from the ground. Falling on the panel, the hailstones leave an imprint which can be analysed when the panel is removed after the hailstorm.

Hailstone diameter	Theoretical absolute Cm_{ij}	Relative $Cm_{(h)ij}$
5 - 10 mm	1	0,2
11 – 15 mm	2,33	0,47
16 – 20 mm	3,67	0,73
> 20 mm	5	1,00

Fig. 4.13 Magnitude-coefficient values for hail sub-risk

Values for relative areas will be obtained by considering the strongest phenomenon with a higher frequency than one single episode (i.e. 2 or more episodes).

4.2.5.4 Storms and Thunderbolts

It is a characteristic of storms that there are various accompanying meteo situations. At times, a single meteo situation may be persistent but its effects cannot be grouped together in the time-sequence of the phenomenon itself. Storm features are: relatively short time of rainfall, from one to three hours, if the rain's intensities are measured; limited space covered; strong, fast variations in pressure and temperature; abrupt changes in wind-force and direction; high concentration of thunderbolts.

The hazard of these situations lies in the simultaneous presence of accompanying factors which potentially can affect buildings. In particular, a storm carries with it a more or less intense barrage of thunderbolts. These can be destructive for buildings with no lightening-protection. The bolts' hazard level is normally higher than other phenomena. This is why we have combined the valuation of storm features with the eventual added problem of high values for thunderbolt-density, i.e. the yearly number of bolts striking the ground N_g.

The formation of classes of intensity for storms can be done by using the list of past series recording intensity and frequency (number of annual storm-days T_d).

If no data are available, storm-intensity classes can be formed through storm-cause models and using the valuation of the TT Index (Total Totals Index), with functions of pressure, temperature distribution and atmosphere-humidity values—in this matter, too, data, though often approximate, can be had from the meteo stations.

Figure 4.14 shows values for magnitude coefficients of the storm-frequency component, $Cm_{(h)ij}(T)$. These values, as we will explain, have to be added to thunderbolt-component.

In Fig. 4.15, the magnitude-coefficient values for the thunderbolt-density component to add to the storm-frequency component values.

On the basis of the data obtained for both components, referring them (already done) to the corresponding areas or polygons one can insert the storm—and bolt fall—into (4.6), using those $Cm_{(h)ij}$ produced by the simple rapport:

$$Cm_{(h)ij} = Cm_{(h)ij}(F) + Cm_{(h)ij}(T).$$

Frequency significant storms (days per year - T_d)	Theoretical absolute Cm_{ij}	Relative $Cm_{(h)ij}(T)$
Low (T_d < 12)	1	0,1
Medium-low (T_d: 12-17)	2	0,2
Medium (average (T_d: 18-23)	3	0,3
Medium-high (T_d: 24–30)	4	0,4
High (T_d>30)	5	0,5

Fig. 4.14 Magnitude-coefficient values for storm and thunderbolt sub-risk (storm component)

Thunderbolt density (thunderbolts/year x km^2 -Ng)	Theoretical absolute Cm_{ij}	Relative $Cm_{(h)ij}(F)$
N_f<2,0 bolts/yearkm^2	1	0,1
N_f<3,0 b/ykm^2	1,8	0,18
N_f<4,0 b/ykm^2	2,6	0,26
N_f<5,0 b/ykm^2	3,4	0,34
N_f<6,0 b/ykm^2	4,2	0,42
N_f>6,0 b/ykm^2	5	0,5

Fig. 4.15 Magnitude-coefficient values for storm and bolt sub-risk (thunderbolt component)

The division made in obtaining factors for the $Cm_{(h)ij}$ results for this risk enables us, in synthesis, to devote more attention, at an equal storm-frequency level, to spells of intense thunderbolt-fall.[19] Note that if we have different areas ($S_{(h)ij}$) for storm—and thunderbolt—phenomena such as when using iso-thunderbolt line-mapping, we have to make a slight, intuitive modification to the (4.6) calculation. In this modification, both subcomponents must be considered separately, with weighting equal to half that of the other components of meteorological risk.

4.2.5.5 Ice and Freezing

When ground temperatures fall to zero or below, hazard is created by the change from water into ice, liquid into solid. This change may cause damage due to the process' volume-expansion or to the weight accumulated in passing the thermal threshold—caused by the water's fluid-dynamic properties being annulled. Typically this event makes for problems in unlagged water-pipes and also causes damage from falling or sliding ice caused by sudden changes in temperature.

[19] Data for *number of thunderbolts falling* per year (N_g), and number of storm-days per year (T_d), are correlated in the closely-related figures for different regions of the world and for geographical context (orography, distance from seas, etc.), on the basis of the $N_g = a(T_d)^b$ relationship, where the a and b are also significantly variable.

Annual frequency of days with $t_{min} \leq 0°$ C	Theoretical absolute Cm_{ij}	Relative $Cm_{(h)ij}$
Low (< 3%)	1	0,2
Medium-low (4%-9%)	2	0,4
Medium (average) (10%-18%)	3	0,6
Medium-high (19%-25%)	4	0,8
High (> 25%)	5	1,00

Fig. 4.16 Magnitude-coefficient values for ice and freezing sub-risk

The number and distribution of meteo-stations in Italy is fortunately quite widespread. As a result, one can, in the above cases, arrive at a classification on the basis of the temperature data collected. The latter is based on the frequency of days at or below 0 °C (so-called "frozen days"). For (4.6) calculation, we can use the various Cm_{ij} from the table (Fig. 4.16).

4.2.6 Risk of Interface Forest-Fires

Hazard bound up with the effects of fire in woods or forests reaching buildings depends on the *interface* between an urban zone and a wooded area. Hence, the high-temperatures, incandescent matter or the flames can reach a building setting some of its parts or components on fire.

A recurrent feature of forest-fires, which can become interface threats to buildings, is the fact that very many of them are started, accidentally or not, by human beings.

In view of such considerations, the interface-forest-fire risk hazard was obtained, first of all, by calculating the extensions of areas that were *wooded*, naturally *non-wooded* (or deforested) and/or *agricultural*. Our concern then was the interfacing belt between the above types of land and built-up areas—working in land-strips 50 and 25 ms wide,[20] along the interface lines, with the specifically devised GIS software. On the base of this surface classification, the Cm_{ij} as in Fig. 4.17 were assigned.

The second move for calculating the $C_{HS(wooded\ interface)}$ was to take account of the geographical data for the percentage of areas (attacked by fire in the programme's reference territory) in relation to the whole of Italy's territory (and in the same manner, it might be used in other countries). And such data were provided with ever growing precision by the State Forest Service, the Regions and local bodies. Working like this, we can consider not only past tendencies, such as criminal interference and arson in the territory, but also the territory's varied climate which may influence soil humidity or weaken plant-life making the zone

[20] Cf Operational Manual for setting up a commune or inter-commune Plan of Civil Protection—Interface Forest-Fire Risk, DPC-PCM, October 2007 (ordinance 3606, 28th August 2007).

Fig. 4.17 Magnitude-coefficient values for forest-fire interface risk

Woodland fire risk	Cm_{ij}
Natural non-wooded or agricultural surface	1
Non-wooded interface surface	2
Wooded surface	3
Thinly wooded interface surface	4
Thickly wooded interface surface	5

more vulnerable to the spread of flames. Calculation of weighting i for interface forest-fire can be made with this equation:

$$p_{i(wooded\ interface)} = \frac{\left(\sum_{j=1}^{n} S_{ij} \times Cm_{ij}\right)}{S_{tot}} \times \left(S_{prov-burned}/S_{naz-burned}\right), \qquad (4.7)$$

More refined classification of Cm_{ij} will be made, if data are available. Sometimes, in fact, there are data readily available for territory areas with different types of plants and crops, which are occasionally available, their vulnerability and propensity to fire upgraded data. On the basis of our test simulations, this information is approximate and estimated, as regards a final result, generally of a secondary level.

4.2.7 *Hydrogeological Risk: Landslides*

Calculation of landslides as hydrogeological hazard began with the data available from our research in the Lecco Province. The data were supplied by the IFFI project (Inventory of Landslides Phenomena in Italy) of ISPRA[21] in collaboration with the Italian Regions and Autonomous Provinces (Triglia et al. 2007). These data represent the first homogeneous and up-to-date landslide-inventory, covering the whole of Italy's territory. So they can serve for applying procedures to the whole country. The IFFI data is supplied through a special SIT, Territory Information System providing the *shapefiles* for GIS.

Information from the IFFI project which is useful in defining territorial hazard regards: land use, kinematics, lithology and, partly, soil condition (for linear landslides). Specifically, the analysis of landslide disruption in Lombardy (Carelli-Ceriani-Valluzzi 2006) resulted in an Index of Landslide possibility % (IF) resulting from the relationship between the total landslide area and the whole mountain/hill area seen in two ways: the *lithological classes* present and the *land use*. As well as such data, information is offered on types of kinematics and mechanisms present. For linear landslides, we are also offered corresponding activity situations, which mean the various Cm_{ij} can be assigned for calculating landslide weightings, as defined in Fig. 4.18a, b and c.

[21] http://www.isprambiente.gov.it/it/progetti/iffi-inventario-dei-fenomeni-franosi-in-italia.

(a)

Landslide type	Cm_{ij}
Collapses, subsidences and widespread instability	2
Slopes with deep gravitational deformations	5
Conoids	2
Linear landslides – active/reactivated/suspended fast	5
Linear landslides – active/reactivated/suspended slow	4
Linear landslides – generally quiescent	3
Linear landslides – stabilised	1
Linear landslides – undetermined	4

(b)

Lithological classes	Cm_{ij}
Diorite/gabbro/serpentinite	5
Landslide debris	4,5
Metamorphic rocks and argillites	4
Sedimentary rocks	2,5
Clays and sands	1
Not available	4

(c)

Land use	Cm_{ij}
Vegetation scarce or absent	5
Grass and/or shrubs	4,5
Permanent cultivations	3
Wooded zone	2,5
Meadows	2
Mining zones, rubbish dumps	1,5
Varied agriculture zones	1,5
Built-up zones	1
Continental waters	1
Commercial zones and communication networks	1
Arable land	1

Fig. 4.18 Magnitude-coefficient values for **a** Landslide risk for *Landslide type*, **b** Landslides for Lithological *classes* and **c** Landslide risk for *Land use*

The calculation of weighting *i* for landslides, since it is done by taking account of indices from historical analysis of data, i.e. analysis of data from past occurrences, will also use the $C_{HS(landslides)}$-coefficient as a single unit. It can be calculated with the equation below, where one notes that, for practical reasons, the average for the three components mapped across the whole reference territory of the programme (landslide types, stone classes and land use) has been arrived at by giving each component an equivalent weighting, dividing a total triple area S_{tot}.

$$p_{i(landslides)} = \frac{\sum_{j=1}^{n} S_{ij} \times Cm_{ij}}{3S_{tot}}. \qquad (4.8)$$

4.2.8 Hydrogeological Risk: Avalanches

In covering territorial hazard of such a level as to damage buildings[22] by avalanche (component of hydrogeological risk), it was not possible to start from thematic mapping. In Italy the maps are available but they are not homogeneous in method and coverage—in spite of excellent work done in Italy by AINEVA, Associazione Interregionale Neve e Valanghe (Inter-regional Association Snow and Avalanches) on the CLPV, Carte di Localizzazione Probabile delle Valanghe[23] (Maps for Probable-Avalanche Localisation). These LPV maps, in fact, show zones with *high, moderate*, and *low* avalanche threat in *red*, *blue* and *yellow* zones. Such divisions are based on estimated return period of phenomena and hypothetical avalanche-pressure data (measured in kPa).

Consequently the first geo-morphological feature to consider is the *altitude distribution* of the reference territory. Altitudes higher than 750 m[24] above sea level ("asl") represent the minimum condition in Italy for avalanche-risk. This is in view of permanence of snow-cover, thickness of snow-drift, the seasonal occurrence of thermal zero, windforce and the proximity of buildings to the snowfall altitude.

The intermediate altitude that means greater risk, considering the above features, is 1,500 m above sea level (asl); above 3,000 m the risk is real—but it is rare for an avalanche set off at this height to have enough energy left (in its descent) to get as far as inhabited locations. Slopes of more than 30° are a requisite for a descent at speed, which spells sudden hazard for buildings. The forming of avalanches also depends on slope's surface, particularly when there is no wooded coverage, which is yet another factor in avalanche-risk. With the data available, one can define Cm_{ij} coefficients for avalanche-risk hazard as shown in Fig. 4.19 below. There, we take as valid the data for the same zones already used in defining the *snow-risk* (so for other areas, the altitude values will have to be adapted).

Though quite often avalanches are caused by careless tourists or skiers, we do not intend to use such cases taken from local accident files. In fact, geo-morphological information should replace data from such accidents—for which, in any case, data are not already available, in view of this, the formula for *i* weighting for avalanches, is simply as follows, taking $C_{HS(avalanches)}$ as one unit:

$$p_{i(avalanches)} = \frac{\sum_{j=1}^{n} S_{ij} \times Cm_{ij}}{S_{tot}}. \tag{4.9}$$

---

22 Note that taking the buildings as targets means we are maintaining the aims of our certification, which is *not* concerned with hazards for tourists or skiers.

23 23rd February 1999 marks the latest death in a home (Dailley, Valle d'Aosta). In the same year, in the Alps, at Galtür (Austria), an avalanche smashed into parts the village, causing 31 deaths.

24 Differently from (*Bignami* 2012), where the altitude was 800 m asl and the 3,000 m threshold was not present.

Avalanche risk	Cm_{ij}
Unwooded areas (1500m<altitude<3000m; incl. >30°)	5
Unwooded areas (750m<altitude<1500m; incl.>30°)	4
Wooded areas (1500m<altitude<3000m; incl. >30°)	3
Wooded areas (750m<altitude<1500m; incl. >30°)	2
Areas at altitude>3000m with temp. incl . >30°	1

Fig. 4.19 Magnitude-coefficient values for avalanche risk

For Verbano-Cusio-Ossola we have also considered slope thresholds of 40°, (which is another, more hazardous threshold used by AINEVA,[25] beyond the 30° one) as well as the distinction between thickly and thinly wooded, the north and south-east exposures of slopes, and snowfall data. This was an effort to produce a more detailed classification. We must admit, however, that this effort has not led to any notable improvement aiming at obtaining a weighting parameter for Italian Provinces related to risk analysis, i.e. considering how much effort went into the data search, classification and the calculation that had to be made. If, instead, we have at our disposal the "Map of Probable Avalanche Localisation" and/or the phenomena covered in the Registers of Avalanches or in the Maps of Zones Exposed to Avalanches (PZEV = Piani delle Zone Esposte alle Valanghe; Barbolini-Natale 2005), just as it was done for hydraulic hazard and for hydrogeological hazard with landslides, we could give magnitude coefficients to specific areas. At the moment, it is not possible in Italy, because of the rather infrequent availability of this kind of maps, but in other countries it could be easier, as, for instance, in Switzerland, where official guidelines on the definition and production of avalanche hazard zones exist since 1984 (BFF/SLF 1984). This would also lead to a past-incidence coefficient $C_{HS(avalanches)}$—if necessary, too, taking the possible different types of avalanche i.e. with *loose snow* avalanches, *slabs* avalanches, *powder snow* avalanches (that can be considered *mixed avalanches*, after some classifications) and *wet snow* avalanches.

4.3 Audit-Standard Plan

Once the three phases of activity for initial assignment of territorial hazard evaluation have finished, the programme in progress splits into different audit plans according to the "standards" foreseen by CEPRED, (see Fig. 4.1), for *each single building in the programme*.

[25] http://www.aineva.it/.

The moves provided for in the audit plan are made with the support of the relevant checklists and calculations, and also with the support of mentioned PREDICT software, concerning territorial hazard and building's vulnerability (sometimes divided into various sections)—both external and internal risks are dealt with. We now go into these different phases.

4.3.1 Desk Analysis of Site Local Hazard

Phase 4, the start of any *audit plan* is known as *preliminary hazard analysis of site*. It is carried out as a desk approach, for each building in the programme. It intends, on the local scale, to examine and verify all the information collected during the start of the programme-definition.

It is a question of investigating, for each risk, all the phenomena and possible disaster-intensities that could affect the building undergoing audit procedures. To this end, following the CEPRED scheme, (the *Schedule for valuing buildings'-hazard indices*), all the data collected during the initial inquiries for the entire audit programme that has been initiated.

The specific aim of this phase is to gather the first useful data for the H_{iL} hazard level. This applies for every external risk i present in the programme territory and in the building's specific location. In this way, too, the auditors are enabled to make their first acquaintance with the building's geographical site, which is essential for good management of the audit's remaining steps.

4.3.2 Preliminary Document-Analysis

Preliminary document-analysis is to ascertain that the documents required by current laws for a building conform to rule and are technically adequate. This first step in analysis may take place at the start of the inspection visit or before this visit (since it needn't be in the building itself). This previous check may, for example, be useful in the case of public buildings belonging to an organisation with various offices in the territory.

Clearly, this documentation can be anticipated, remembering also that it follows a consignment of an *information prospectus* listing in detail the required documents. Of these, the CEPRED do not keep copies but only take written note of them—unless there are situations, in agreement with the client, that need special investigation.

Phase 5 involves collecting more data about the building. It can also mean getting the first results for certification. This would be the case, if there turned out to be non-conformities, especially regarding the already-mentioned *red inquiries*. If the responses were not satisfactory, it would mean annulling the certification whatever the other audit results might be. It would mean that the building gets a *blue shield striped with red.* A first requirement for CEPRED's purposes is that the building under examination is not under any restructuring or, more generally, that no building work of any kind is going on or that no DIA (or similar procedure) is active (in Italian: Dichiarazione di Inizio Attività = Declaration of Commencement of Activities) or other similar procedures in progress.

Phase 5 is run with the help of a vulnerability checklist, *Form for valuing buildings' vulnerability indices*, to complete in double copy—the auditors, as we said before, are, at least, two in number. As regards preliminary document analysis, the information prospectus which had been sent to the client so as to ensure best results in the verifying procedure and the following data-collection, also reminds to find or recuperate as completely as possible documents and information (note that some of the requests do not apply to dwellings with only one family):

- Contract or concession;
- Permits (licence for building) for the current condition of the building;
- Declaration of habitability fitness and fit for use (and/or equivalent document);
- Fire Prevention Certificate or Provisional Permit issued by the Fire Departments;
- Gas system certificate;
- Electricity system certificate;
- Heating or conditioning system certificate;
- Radio-tv and electronic devices certificate;
- Geotechnical analysis of building's landbase;
- Lift systems documentation (for people and goods);
- Connection to municipal sewage network documentation;
- Construction year and eventual important building's refurbishment records;
- Presence in the building of cooking surfaces (with or without electric switches) or devices for fuel stops supply if flames are to be extinguished;
- Date of setting up and type of gas system and its features (presence of flexibles, rubber tubes);
- Instructions safety (in Italy followed as per law 81, 9th April 2008) regarding workplace in blocks of flats;
- Maintenance state of anti-fire devices (extinguishers, hydrants, etc.) and fire-fighting tugs (gloves, masks, helmets …).

Obviously, this list having been thought for Italy, it has to be re-examined and/or integrated in each different national context on the basis of local rules in force (or of better standards).

The audit proceeds with the building's inspection visit, which has been arranged beforehand with the owners or their reps. If the building offers particular problems, there may be, prior to the visit, a detailed specific plan of the building's audit sent to the owner so as to allow or guarantee that he or his rep can be present at convenient time. This is also to be sure of access to different parts of the building or to neighbouring properties, thus limiting any waste of time and also avoiding interruption of, for example, business activities.

4.3.3 Direct Field Check on Hazards

Phase 6 of the *inspection visit* starts with a *direct field check on local hazard* with direct contact on the site, noting the scale of the building to be valued and its surroundings. Inspection has to be made for all the risk present in the programme's reference territory, which have been noted during desk analysis for local territory hazard (phase 4) or have possibly been found during the inspection itself. So, in fact, the direct visit also provides a check on the reliability of the *collected documents* and on the accuracy of the statement of the structure's condition (if not, the PRED process goes back to *phase 4*).

After checking the documentation, the inquiry into the site dangerousness, goes on to seek traces of past events through eventual *interviews* and analysis of historical archives (or their equivalent). This research becomes significant if there are signs that particular details have not been included in the documents collected in the programme scale for different risks.

At the same time, this site research helps in gathering some information on the building's *context*. These data are needed for valuing the *context vulnerability*, V_C. This is a component of the later vulnerability search, and is especially useful for estimating internal risk—as will be detailed in the following paragraphs.

The next step is to see, inside the building, whether there are means of protection from special risks or monitor and alarm systems installed after archived events and therefore modifying localised hazard for the better. A similar valuation is made, if information is available, for "foreseeable" phenomena so as to evaluate local *emergency planning* and any specific self-protection measures applying to single building's or inhabited nuclei. Such data also serve for valuing vulnerability. Take for instance the techniques of *floodproofing* (cf. Rosso-Bocchiola 2006), the presence of means for *woodland firefighting*, that of machines for *snow* removal and the existence of protection safety barriers to limit the consequences of industrial disasters (cf. Lénoble-Durand 2010), etc. Consider these means together with a valuation of the chances of an efficient and timely intervention by personnel who are trained and technically capable. From a logical point of view, to consider this kind of intervention poses a special issue, because it concerns activities operating in real time, during the disaster dynamic evolution just before reaching the maximum of its effects intensity. From a mathematical point of view, it considers how risk (or *specific risk*) is as a function not only of hazard, vulnerability and worth exposed (or only of hazard and vulnerability), as shown in the (2.1) (or in the (2.2)), but also a function of *coping capacity* C, as in the theoretical formula below:

$$R(E,I) = \frac{H(I)V(I,E)W(E)}{C}. \tag{4.10}$$

It is immediately evident that to comprehend, that with C growing, risk is reduced. In spite of all this, measuring coping capacity is unfortunately a hard work. That is the reason why we do not use the (4.10), but we include this capability of intervention as a component reducing vulnerability (cf. Mojtahed et al. 2012).

In the case of buildings not recently constructed, the attempt is made to value their eventual maintenance condition. To value hazard levels H_{iL} (or, analogously, $H_{L(i)}$) for each separate external risk i, we use procedures, that are similar or different, depending on the precise features of the various possible phenomena.

This phase finishes the *checklist* which was started in *phase 4* of the site's desk analysis for hazard (the *Form for valuing buildings' hazard-indices*).

4.3.4 Synthesis of Site-Hazard Valuations

As already mentioned, the book here offered to the readers does not aim at being a manual[26] for applying CEPRED (this being a research work, propaedeutic to an executive stage) but it aims to show, overall, that it can be realized—what we show here is the procedure to be used to obtain the hazard component in valuing external risks. So we begin by defining the event scenarios that could strike the buildings being valued.

- The *seismic-local hazard level* $H_{L(seismic)}$ is obtained starting with the value for maximum expected acceleration[27] in the zone or subzone where the building stands and applying coefficients to this value, taken by the seismic micro-zoning. These coefficients are, respectively, of subsurface categories (C_s) and amplifying topographic factors (C_t) in rocklayers and geomorphics such as crest *slopes* and *widths*. The value thus obtained is placed in relation to the value for maximum horizontal acceleration in zone 1 (0.35 ag/g). If no info is available on subsurface categories (on the *basis of equivalent spread-speed for cutting waves* V_s 30 m/s), we give this coefficient the prudent value of 1.25.

$$H_{L(sismico)} = \frac{a_{\max \times C_s \times C_t}}{0.35}. \tag{4.11}$$

- The *level of local hydraulic hazard* is obtained from a more detailed coverage of the building's position as per the natural catchment area and the man-made water-network. This procedure involves questionnaires and scores deriving from the EU interreg project *Quarter*.[28] For valuing flood-risk, these inquiries investigate many different aspects which we can sum up as follows:

---

[26] Obviously, such a manual could be compiled and distributed as soon as the spread of certification makes it necessary for authorised skilled workers.

[27] This represents the *basic seismic hazard* of building sites as foreseen by the Italian technical norms.

[28] Technical attachment for valuing flood risk (E. Larcan, S. Mambretti with D. Bagioli, M. Cameroni, March 2004).

- eventual position in fluvial or flood prone-areas;
- structures (standing alongside water courses) that can have an effect on a building;
- the presence of bridges;
- eventual geological studies about the character of the terrain in question;
- the existence of maps showing damage from past events;
- local studies on high-water levels and overflows in the area;
- the general territory-morphology (flat or with hillocks, with general ups and downs, or sunken areas);
- the maintenance condition of any present riverbanking structures; vegetation under control; formation of blow-hole;
- how long it takes for overflows to recede below main-bank top.

Beside the above, when possible with the available data, we added consideration of information on foreseen water levels and on overflow speeds and its combination—in view of the thrust they can exert on built structures.[29] In fact, some phenomena may not be significant in compromising a building's resistance, if the values for water level and flow are acceptable. The flow speeds used in CEPRED are 1.0, 1.5 and 2.0 m/s and thresholds for water level (in the angle between ground and building's outer wall): 0.30, 0.50 and 0.70 m. Instead of speed values, if not available, one can use slope-inclination values, if they are significant, considering local morphology, eventual urban context and the surface roughness coefficients (such as Gauckler-Strickler roughness coefficient, for instance).

The hazard points given also take account of the contribution of two factors, on the basis of an inversely proportional combination as acceptability threshold depending on a fixed interval of forces (cf. Rosso 2002 and 2005).

The hazard values $H_{L(hydraulic)}$, for their final results, will be in conjunction with the data from the building's structure (as in vulnerability valuation) and any eventual floodproofing measures taken in real-time.

If a building should turn out to be in a zone where, according to the last laws or decrees, building is not allowed (areas of regard, areas with high or very high hydrogeological risk, areas A and B of Italian PAI indications), the hazard valuation for hydraulic risk can come directly to a negative conclusion, assigning a *red-striped blue shield.*[30]

- The calculation of the *chemical-industrial local hazard level* $H_{L(chemical\text{-}industrial)}$ is mainly based on coefficients assigned according to the distance from the

[29] The main damage mechanisms for buildings from flood concern: *hydrostatic actions*, such as *horizontal pressure* and *capillary return*; hydrodynamic actions linked to waves and to water flow-rate (turbulence included); erosion and undermining of the buried parts of building; *floating* thrust; damage from solids-transport material; chemical damage from transported contaminants (Kelman and Spence 2004). Most of these mechanisms depend for their effect on the water's depth and speed.

[30] Generally, these red-striped shields are given for hazard valuation in view of criteria regarding how unpredictable events are, their intensity and frequent recurrence.

danger-centre (installation at high risk), the presence of eventual natural or man-made barriers, the eventual position in zones covered on External Emergency Plan zone 1, "*lethal effects*", zone 2 "*serious injuries*" and zone 3 "*reversible effects*") and by the related land-use restriction zones. Different points are given for these distance thresholds and zones since they are distinguished depending on the possible dynamics of an event: *instantaneous* (e.g. explosion of fireball), *hours-long* (such as heat dispersion, toxic fumes, fires out of control) or *days-long* (spilling or release of toxic products on the ground or into water courses. Both of the methods present in literature, *consequence-based* (deterministic) and *risk based* (probabilistic), are considered and included in our procedure (cf. Christou et al. 2006). Next programmed investigation steps on *chemical-industrial dangerousness* will be the inclusion of the evaluation of the potential domino effect, also taking into consideration natural disaster triggered by industrial accidents.

Since each plant may produce different scenarios and, as there may be more than a single plant involved and since the accidents must be treated as *independent* from each other and not *mutually exclusive*, calculation comes about through a formula giving the $H_{L(chemical\text{-}industrial)}$ value according to the following criteria (including *common occurrence*):

$$H_{L(chemical-industrial)} = 1 - \prod_{s=1}^{k} (1 - H_{Ls}), \qquad (4.12)$$

where the values for dangerousness H_{Ls} for each of the k scenarios s must be inserted.

If a structure comes under a zone of sure impact or damage (excepting spillage or release on ground or into water), then, independently of the points that may have been gained in the whole audit-procedure, the zone will receive a *red-striped blue shield*, which means any positive certification is out of the question.

- Calculation of the *local hazard level for transport risk*—transportation of dangerous goods as well as proximity to transportation routes $H_{L(transports)}$ takes into account both the components for direct-impact damage (including plane crashes) and damage from dangerous-goods traffic. Each possible scenario comes under a formula like (4.12). Reconstruction of these scenarios for nearness to transport ways includes the physical features of the actual transport ways, the building's placing with regard to the latter (distance, nearness to the road-bends, overhead roads, etc.) and refers to the building's response to eventual impact[31] from vehicles. For the transport of dangerous goods information is obtained as to the frequency and location of past accidents and on the quantity of dangerous goods in transit.

[31] Technical norms define like this actions taking place only exceptionally in the expected lifetime of a building.

Certification is nullified and the red-striped blue shield is presented when dangerous goods are passing within 50 m from motorways or state roads (if the latter have a speed limit exceeding 90 km/h) or from railways, gas or oil pipelines—in the latter cases precautionary values were given when the commune authorities were unable to define the exact location.

- *Meteo local hazard level* $H_{L(meteo)}$ brings together the hazard of the 5 sub-risks (weighted equivalently), giving each of them a value for the building's geographic zone in which events occur with more or less intensity and/or frequency (on the basis of the mentioned meteorological events' scales and on the basis of the local morphology and eventual special condition of the buildings sites, if they are significant).
- The *interface forest-fire local hazard level* $H_{L(forest\text{-}fire)}$, in a deep belt stretching 200 m from the building's perimeter,[32] takes into account the various factors that can lead to contact between a building and a forest-fire or, in any case, a wildfire out of control running across the area with abundant vegetation. These factors include the type of vegetation—more or less quickly consumed by flames on account of the mass/surface rapport, or the presence of undergrowth etc. They also cover the vegetation-density, the slopes of the area around the building, the contact, whether more or less continuous between wooded or open areas, and the building's perimeter, the local climate characteristics of the sites, eventual past fires or nearness of water bodies and possible further classifications present in local emergency plans.
- *Hydrogeological local hazard level from landslides* $H_{L(landslides)}$ considers, first of all, whether the building stands on areas of landslide territory within its surrounding belt 100 m-wide—not forgetting the inevitable uncertainty natural to landslide phenomena. And if a building proves to be directly involved in a disaster area (by an inquiring as whether there are detailed geological maps of the zone or by direct investigations on the field), it will receive the *red-striped blue shield*. If this is not the case, then the distance from the disaster area will be noted together with (if known) the advancing speed of the phenomenon.
- For *hydrogeological local hazard level from avalanches*, $H_{L(avalanches)}$, the first inquiring is whether there are documents regarding the zone, such as maps of Probable Avalanche Location, or Avalanche Registers, or Plans of Zones exposed to Avalanches or whether there are traces, even recent ones, of past avalanche phenomena affecting the building and its surroundings. A second approach is concerned with the building's above-sea-level position, mountain-slopes or hill-slopes, possible woodland, the slopes' exposure, the slopes' stability (possible landslide-zones), and the eventual presence of defence structures (supporting structures). As in the previous cases, unless later defence structures

[32] Cf Operation Manual for a commune or inter-commune Plan for Civil Protection—Interface-Fire Risk, DPC-PCM, October 2007 (ordinance 3606 August 2007).

have been put in place (even temporarily in case of simultaneous combination with reforestation measures cf. FOEN/WSL 2007), the insertion of the building in red or blue PZEV zones and the building's involvement in past avalanches or the involvement of its location area, if the building is a later construction, means the immediate presentation of a *red-striped blue shield.*

4.3.5 Direct (Visual) Checks on Building

After the preliminary documentary analysis and the inquiry into the site's hazard levels, we come to *phase 7* of our check. This phase means to value the building's physical situation—as regards its internal components and its outer shell. The intention is to evaluate the different V_{iL} factors for each of the building's risks, the *general vulnerability* V_G, the *context vulnerability* V_C and the situations for the various *internal risks* $R_{int(k)}$. The check on this phase also includes use of the *Form for valuing buildings' vulnerability indices.*

Detailed procedure for estimating vulnerability and valuing the safety level regarding internal risks started out with the self-valuation form "*Prevention of collapse risk in residential buildings*" developed by the Milan Polytechnic for the Lombardy Region. This product was much appreciated but, after further reflection and later testing, some modifications was made to it so as to widen its content and adaptability to chime in with CEPRED's aims. It was also necessary to cover some further risks under examination, such as, for instance, a building's reaction to seismic tremors, for which we had recourse to the *form for first-level damage report, emergency intervention and fitness test for ordinary buildings in past-seismic emergency* (AeDES). This form had been invented by the Civil Protection Department and the National Group for Earthquake Defence, above all for the purpose to be used after disastrous earthquakes in Italy. But, in our applications, it has also shown its effectiveness as a prevention support tool.

4.3.6 Building's Vulnerability

The scheme for analysing vulnerability from external risks started from the approach to be found in the *Quarter* project, which was then adapted to include some new risks and integrations. Some questions were formulated to extend to some types of possible phenomena and ascertain the *interaction* between them and the building's structure and features. This was to value resistance or giving way to load action.

- In particular, as for collection of information and data for valuing *general vulnerability*, the intention is to verify any basic tendency of the building to suffer damage—without referring to any specific risk when facing any kind of danger. Thus, the analysis points to: *legal conformity*, physical jeopardy due to *transformation, deterioration* or *breakdown*; features linked to special usage and goods with historical, artistic or particular value economically or technologically. General vulnerability is a term used in the final calculation of the *Index of edifice's resistance predisposition to external risks* $I_{res(ext)}$ on the basis of (2.3); this takes on a value greater than or equal to 1 (up to a maximum of 1.9), fulfilling an amplifying function in the result for the weighted aggregation of the various specific risks $R_{s(i)}$ for each of n external risks covered by the activated audit programme.

- The basic treatment for calculating the various *vulnerabilities* (or various *vulnerability levels*) V_{iL}, for each external risk, depends on parameters valued overall from the replies to the above-mentioned checklists. This information comes particularly from the list answers on hazard, containing the main data about the presence and intensity on the risks in the programme.

The main parameters considered for each risk, valued by its own specific phenomenon, are as follows:

- *Reachability*—parameter (after an event, or in some foreseeable scenarios, before an imminent event) to ascertain whether and how a building can be reached through the road network.
- *Accessibility*—differently from the above, this parameter values the possibility of the helpers entering or exiting a building after a calamity—if necessary using emergency means (firemen's ladders, etc.).
- *Physical jeopardy*—one of the most important parameters. For each possible phenomenon, it values the parts of a building that are able (predisposed) to resist the intensities of possible events.
- *Functional jeopardy*—this values the technical function present in the building and the special uses in specific rooms such as, for instance, computers or electronic devices working for medical application or house-management.
- *Jeopardy for eventual relevant devices*—in this case, the concern with technical or technological devices which may also be working for customers not in the building itself (for example, workshop devices or commercial or professional services).
- *Jeopardy for culture objects*—the Art Inheritance, movable or immovable, is, without doubt, pervasive in Italian society (and abroad, especially in European and Asian countries). This parameter[33] values the more or less noteworthy presence and vulnerability of art objects—both as regards a building itself and the valuable objects inside it.
- *Functional autonomy for electricity, water, communication, sewerage system, energy supply*—these parameters deal with autonomy of technical services, valuing for example, the positional number of meters and electrical systems,

[33] An eventual expansion of CEPRED in future might provide a separated valuation procedure for edifices of historical and architectural interest.

types of connections and providers, presence of generators, connections to aqueducts, presence of autoclaves, boilers and/or gas cylinders or combustible materials, telecommunication systems, etc.

- A role similar to that of general vulnerability is that played by detailed information on existing documents regarding the building's original planning and construction and documents concerning the building's nearness to other worksites. The latter might be recent nearby excavations (for garages, for the underground, etc.). Here, we must add information on the observance and updating of norms for *health* and *accident prevention* inside the public building. This might include, for example, gymnasiums, school laboratories, and even banal things like anti-slip strips on stairs where quick evacuation has to be foreseen. All these types of information were gathered in detail for the specific kinds of building involved. The first stage was a series of questions for single-family dwelling. Then, there was a gradual increase in the number of questions for condominium buildings (which are subject, in Italy, to particular regulations according to the D. Lgs. 81/2008), for office buildings and schools. Each of these check—and inquiry-groups, which point to *indirect indicators* of a building's vulnerability, are given points and become aggregated after normalisation producing an average value called as *vulnerability of context* V_C.

This term is used in the final calculation for the *Index of edifice's resistance predisposition to internal risks* $I_{res(int)}$ on the basis of (2.5) and we obtain the following formula:

$$V_C = \frac{\sum_{g=1}^{3} \sum_{i=1}^{n} P_{Q(g)i}}{3}, \tag{4.13}$$

where the $PQ_{(g)i}$ are the scores obtained in the various n inquiries i about each of the identified g ambits grouping indirect indicators that are useful in valuing the internal-risk context (*existing documents, nearness context, accident prevention/ health protection*).

4.3.7 Levels of Internal Safety

For internal-risk level $R_{int(k)}$ (on the basis of the completed questionnaires and the building's scores obtained from them), we use the following simple formula:

$$R_{int(k)} = \sum_{i=1}^{n} P_{Q(k)i}, \tag{4.14}$$

which is useful for calculating (2.5) to obtain the *Index of edifice's resistance predisposition to internal risks* $I_{res(int)}$. In (4.14), the various $P_{Q(k)i}$ are the scores from the different queries i about each of the internal risk k.

Namely and precisely, the three internal risks considered are collapse, fire and explosion. These are the direct factors which have their part in deciding the greater or lesser degree of a building's intrinsic safety without counting whatever happens or is present in its surroundings.

4.3.8 Collapse Risk

As already stated, the approach to this internal risk is based on research on collapse risk carried out in 2001 by the Milan Polytechnic for the 5th Civil Defence Booklet "*Prevention of collapse risk in residential buildings*" published by the Lombardy Region. For CEPRED's purposes, the above work underwent a reorganisation of the already-developed material—in particular to adapt some of its features to the fact that analyses were no longer on a *self-valuation* basis but were now in the hands of a competent specially-formed team. And their results would now bring significant scores or, in the most critical cases, would stop certification.

The stability of a building depends on various factors which, through time, influence its nature. Typical signs of anomalous structure-situations may derive either from technical features, due to later modifications to a completed building (though there may also be deficiencies in the original structure), or from wear and tear on an old building. In the latter case, the time threshold for increased attention is forty years for a building's age. In Italy, this threshold is a cause of worry because of the growing number of houses built after the Second World War, often with suspect quality (not to speak of buildings rebuilt but on top of what was left after bombing). The postwar buildings have received very little maintenance work. Since 1951, in Italy, the total number of dwellings has much more than doubled.

Valuation differs depending on whether a building was made with reinforced concrete or in brick. The queries and checks were designed so as to bring out any hidden faults—why a building might conceal from a careless observer. Logically, this section has been thought bearing in mind characteristics of Italian stock of edifices. To be applied in another country, especially this part of the PRED audit process, will need an integration of questions on the local techniques of construction. For instance, in many countries, especially in the USA, a great number of beautiful and expensive houses and buildings are wooden edifices.

4.3.9 Fire Risk

The treatment of fire risk was developed *ex novo* (fire risk was not included in the mentioned Booklet for Civil Protection, "Prevention of collapse risk in residential buildings") with reference to the Italian official norms. The queries and checks face various aspects, among which, for example, we find: the presence of work-rooms with significant use of fire and heating, features of supporting structures, position of smoke-alarms and fire-alarms, position of fire-fighting devices and systems, presence of personnel trained to use fire-fighting devices, availability of devices for individual protection, presence of fireproof walls and doors, enough practical exits, clearly visible warning and exit signs, evacuation plans, presence of efficient lightening-rods. Just as norms, in Italy, differ for private dwellings for workplaces and for schools (as in many other countries), the queries and checks for certification have been similarly organised. In this case, too, questionnaire would be adapted to the context of the PRED application country.

4.3.10 Explosion Risk

For explosion risk, too, the audit form is based on the already-cited 5th Booklet for Civil Protection, "*Prevention of collapse risk in residential buildings*".

As for collapse risk, the cases have been slightly extended. The biggest problem that had to be tackled was the fact that in the larger condominiums, in Italy, each apartment had a gas system (for instance, for heating and the hot-water boiler) and its risk had to be estimated in combination with others.

The first problem was giving scores in situations where poor quality typified various systems. The solution was overcome with a calculation of *disjoint probabilities*. A different choice, such as giving points according to an inverse linear rapport between the number of unsatisfactory systems and the safety-level score would have led, in our opinion, to unreasonable severity in the valuation.

The second problem was the difficulty of valuing each apartment. Though this is recommended in the procedure, it would be a disadvantage for the larger condominiums. The problem was partly solved by an analysis based on circumstantial factors like the age of the occupiers and their social standing (though always bearing in mind a sensibility for privacy emotional response in presenting the plan of inspections). These factors helped to single out potentially critical and unavoidable situations. For other apartments, however, at least a third[34] of them randomly selected would undergo a field inspection. In this case, i.e. a potential inspection of a building's system, the final result for explosion risk will be limited by a prudential coefficient cutting the figure by 10 %, trying to limit the unavoidable element of uncertainty introduced by working on a sample (even though not so small).

---

[34] On the basis of the norm ISO 19011:2011: "*Audit evidence should be verifiable. It will in general be based on samples of the information available, since an audit is conducted during a finite period of time and with finite resources. An appropriate use of sampling should be applied, since this is closely related to the confidence that can be placed in the audit conclusions*".

4.4 Final Desk Processing of Collected Data

Phase 8 is the moment when all the data gathered through checklists, and, if necessary, the new documentation, collected and selected, are examined and put to use. This means inserting the data into the relevant calculation-programmes (into PREDICT when completed) so as to get results from the various formulae for calculating indices.

Specifically, final values are obtained: for dangerous levels H_{iL} and vulnerability V_{iL} (necessary for calculating the specific risk-levels $R_{s(i)}$ for external risks), for general vulnerability V_G, for context vulnerability V_C and internal-risk levels $R_{int(k)}$. Recalling (2.6), the result of combining indices obtained through (2.3) and (2.5), i.e. the function:

$$I_{res} = f\left[I_{res(int)}\left(V_C, R_{int(k)}\right), I_{res(ext)}\left(V_G, R_{s(i)}\right)\right],$$

which is the *Index of edifice's resistance predisposition to disasters*, representing substantially a *synthetic index of possible safety level* of a building, has so far been obtained, for each of the three experimental contexts (the Milan, Lecco and Verbano-Cusio-Ossola areas), through the following formula:

$$I_{res}\left[I_{res(int)}\left(V_C, R_{int(k)}\right) + I_{res(ext)}\left(V_G, R_{s(i)}\right)\right]/2, \tag{4.15}$$

having simply been smart enough in the maths calculation to normalise the values obtained for internal and external risks by bringing the specific-risk or risk levels back into *safety levels* (a banal 1 or 100 complement operation). The effect was, for communication reasons, that we could assign and print on statement, results indicating low risk and consequently greater safety.

4.4.1 Possible Extra Checks and Eventual Further Re-Examination Audit

The final processing of the collected data intends to show possible critical features that were not directly discovered during inspections. In fact, from this final procedure there may emerge the need to revise or to re-examine the collected data. This may mean the auditors have to make *further checks* or it may even mean *suspending* valuation until the owner clients have completed special extra technical adjustments or examinations. This, at least, would mean the clients becoming aware of the need for extra applications.

Moreover, since for CEPRED, the building valuation is valid for 3 years, before or, when this time expires or, when some building intervention (for good or ill) has taken place or steps have been taken (by local authorities) to reduce the area's hazard or, in any case, if the area has been transformed, then a fresh audit of checking and re-examination can, if necessary, be started so as to review a building's possible performance in resisting disasters.

This re-exam audit will, obviously, also make use of work done before for the first audit, going through all the data already collected and used. This new approach may also signify an *eventual revision* of *territorial classification* on the programme scale—when, for instance new norms are introduced for mapping hazard, when infrastructure is created or abandoned, when a new campaign is under way for collecting new risk data for risk mapping. For sure, a fresh audit will re-examine and redefine the building vulnerability and its site's hazard.

In this way, the PRED certification confirms itself as a proper *safety management system* especially if related to the disaster risk treatment.

Chapter 5
Concluding Considerations: Can We Succeed in Limiting the Consequences of Calamities?

In working for the experimental proposal of a multi-risk assessment certifying process, our approach took on the idea of a *voluntary* system devised to be managed mainly by a third-party body. The system first analyses the *hazards* of building territory and site, *estimates* the *hazard* and *vulnerability* of buildings and, in the case of satisfying positive findings in taking into account possible hazards and vulnerability interactions, eventually produces a *labelling result* for the potential buildings' *resistance* (or strength) in the event of calamity. All this in order to try to shift the building market as a whole towards better *multi-disaster resistance predisposition* performances, but also as an element of endorsement and rewarding the best (5 or 4 *Blue-shields*-rated) properties. The resulting system (at the present, 2013 work conforms itself to the norms ISO 19011:2011, ISO/IEC 17021:2011 and ISO 31000:2009) provides information potentially useful to the various users of our buildings, thanks to an analysis deriving from an objective and reproducible *audit* process. The latter was developed so as to allow for subsequent improvement in *land use planning* and *building regulations on safety*. Such policies would eventually lessen damage to buildings and their inhabitants, showing how disaster risk management could more and more help to protect lives and promote more undisturbed economic development for people and countries.

Thanks to the official tool of the PRED certification (or rating/management system), owners, landlords, residents and renters or buyers will, first of all, find out the specific condition of the building that interests them, since now they have available, as never before, a wealth of detailed information.

This availability has become the basis for initiatives of *bottom-up* protection from natural and man-made disasters—initiatives which may influence individuals in their attitudes towards "self-protection" measures such as advisable building maintenance, or modification and/or purchasing appropriate insurance, as well as influencing official initiatives in correct and prudent *land and building management*—such as wise urban planning in taking care of the land areas and of public safety. The same availability, obtained by means of our more transparent communication and assessment strategies, may also help to enhance *feeling of*

D. F. Bignami, *Towards a Territorial Multi-Disaster Buildings' Resistance Certification*, SpringerBriefs in Environmental Science, DOI: 10.1007/978-88-470-5223-9_5,

responsibility in researchers and professional individuals involved, for example, in setting up and perfecting buildings' design.

Thus, in having permanently in use this system of evaluation and labelling as regards territorial risks and possible damage to buildings there is a double advantage. On the one hand, the system helps the promoting of a *safety culture* for both dwellers and decision makers by encouraging informed awareness on disaster risk. On the other hand, it should, as in the case of other buildings certification, encourage all the individuals involved towards an informed *free choice* in deciding the best measures of improvement, in the case of the PRED necessary for safeguarding their home. Taken together, these mechanisms should attract attention, promote continual widespread discussion on safety levels of built-up areas and considerations about the costs and benefits of possible improvements in the strength and resistance of the buildings in all their components—as well as of achievable enhancements of the general urban settlements' *resilience* (or, as already argued, *adaptive capacity*).

Moreover, such a development would also bring about a more efficient allocation in the building industry of resources concerning safety (both by reducing *externalities* and controlling *land rents*). Furthermore, in the short term, investing in risk management and in risk reduction could be a way of accelerating actions both for growth and jobs opportunity (by the introduction of new technologies and techniques, refurbishments of buildings and infrastructure, innovative financial instruments, research and development, etc.). Likewise, in the medium term, improved disaster assessments and resilience, or adaptive capacity strategies, help to focus on structural sustainability of public and private finances and to improve the macroeconomic stability by reducing the negative impact of disaster on growth and public and private finances. As well as in a long-term perspective, investing in risk management strategies has a high rate of return and is contributing to a more sustainable economic development.

A disaster risk certification, such as that that is offered by this book, could address a number of related policy concerns. It could be operational and synergic at all levels of the risk management cycle, from risk identification (including, for instance, high-resolution hazard maps) and risk modelling to risk control, reduction and response (emergency planning). The proposal of certification can encourage and strengthen habitual strategies for protection of the public and the land but without replacing the old strategies, which could and should be integrated with the new measures. These, more complete, strategies would go together with a body taking widespread initiatives to create greater and long-term respect for the regulations safeguarding the land at every public and private level. This whole body of actions should produce converging effects instrumental to widen and deepen public sensibility, willingness and awareness as to safety.

The final result, narrowing the gap between theory and practice, ought to be that of speeding up the medium-to-long-term *reduction* of human tragedies, economic losses and both those of history and identity that assail the disaster areas. Such an aim is shared verbally by many but its achievement is still ineffectual in method and beyond our grasps.

Considering the complexity of the work of development of the PRED certification here presented, and bearing in mind that the PRED work-scheme application, at least in the case of blocks of flats, will probably require a cost not superior to the sum of the costs of each energy certification statement of flats of the block itself, PRED can be technically considered near to be effectively transposed into operational acts. The PRED certification system, at present, needs only a few more steps to reach an adequate level of trustworthiness and consistency, i.e. to enlarge and refine both the calibration parameter sample and the overall work-flow of the certification (check-lists included); to complete the range of hazard incorporated in the scheme (first of all *volcanic* and *sea-coast* hazard, and adding small sub-cases of hazards to those already integrated by PRED); to develop the software tools to support all the activities involved; to have, eventually, the procedure adopted within each relevant national regulatory framework outside Italy.

In addition, during our calibration campaigns, we have been observing, and keep doing so, that people and communities are positively engaged in the safety of their buildings when seriously and consciously included in due time in the process, as it is in the PRED. This is encouraging as a test, in order to verify social and economic acceptability of the PRED. It must be noted that an important issue to be studied in future is also the fact that the less (or the least) resistant edifices are often purchased or rented by people of means too limited to improve the safety standard of the buildings.

Our aim is to continue and extend this research, investigating large-scale applications also in collaboration with civil protection services, public housing agencies, real estate and insurance companies.

Bibliography

Amantini A (2010/11) Casa Sicura 2012—Prima applicazione alla Provincia del Verbano Cusio Ossola, Elaborato di laurea, Fac di Ing Civile Ambientale e Territoriale (Politecnico di Milano), A.A

Barbat AH, Lagomarsino F, Pujades LG (2006) Vulnerability assessment of dwelling building, chap. 6. In: Oliveira CS, Roca A, Goula X (eds) Assessing and managing earthquake risk. Springer, Dordrecht

Bassi M (2007/08) Progetto Casa sicura:procedura di audit e definizione del sistema di rating per la valutazione della sicurezza territoriale degli edifici pubblici e privati, Tesi di laurea, Fac di Ing Civile Ambientale e Territoriale (Politecnico di Milano), A.A

Basta C, Neuvel JMM, Zlatanova S, Ale B (2007) Risk-maps informing land-use planning processes—a survey on the Netherlands and the United Kingdom recent developments. J Hazard Mater 145:241–249

Bernatik A, Senovsky P, Senovsky M, Rehak D (2013) Territorial risk analysis and mapping. Chem Eng Trans 31

BFF/SLF (1984) Richtlinien zur Berücksichtigung der Lawinengefahr bei raumwirksamen Tätigkeiten. Bundesamt für Forstwesen (BFF) und Eidg Inst für Schnee- und Lawinenforschung (SLF). Eidg Drucksachen- und Materialzentrale (EDMZ), Bern

Bignami DF (2009) Tolosa, 21 settembre 2001—Industria chimica e sicurezza della città, una rincorsa fallita, 550–585, in Città, Salute, Sicurezza—Strumenti di governo e casi di studio—La gestione del rischio, Maggioli

Bignami DF (2010) Protezione civile e riduzione del rischio disastri, Maggioli

Bignami DF (2011) L'iniziativa "Casa Sicura": esiti delle prime applicazioni sperimentali nell'ambito del progetto MIARIA, in MIARIA—Tecnologia e conoscenza al servizio della sicurezza, Bellavite Editore, Missaglia (LC)

Bignami DF (2012) L'iniziativa "Casa Sicura": valutazione degli esiti sperimentali e riflessioni sull potenzialità di diffusione, in MIARIA—Tecnologia e conoscenza al servizio della sicurezza, Vol II, Bellavite Editore, Missaglia (LC)

Bocchiola D, Rosso R (2006) Convivere con rischio inondazione: un'introduzione alle tecniche anti-inondazione, Il progetto sostenibile, 9

Borghesi A, Gaudenzi B (2013) Risk management—how to assess, transfer and comunicate critical risks. Spinger, Italia

Cahen B (2006) Implementation of new legislative measures on industrial risks prevention and control in urban areas. J Hazard Mater 130 (Elsevier):293–299

Carelli M, Ceriani M, Pozza F (2011) Il Piano di Rischio Integrato d'Area per la Valsassina e la Valvarrone, in MIARIA—Tecnologia e conoscenza al servizio della sicurezza, Bellavite Editore, Missaglia (LC)

Cavalli A (2007/8) Progetto "Casa Sicura": analisi multi rischio applicata al territorio della Provincia di Milano, Elaborato di laurea, Fac. di Ing. Civile Ambientale e Territoriale (Politecnico di Milano), A.A

D. F. Bignami, *Towards a Territorial Multi-Disaster Buildings' Resistance Certification*, SpringerBriefs in Environmental Science, DOI: 10.1007/978-88-470-5223-9,

Canuti P, Casagli N (1996) Considerazioni sulla valutazione del rischio di frana, estratto da "Fenomeni franosi e centri abitati", Atti del Convegno di Bologna del 27 Maggio 1994, CNR—GNDCI—Regione Emilia Romagna, pp 57

Christou MD, Struckl M, Biermann T (eds) (2006) Land use planning guidelines in the context of article 12 of the seveso ii directive 96/82/EC as amended by directive 105/2003/EC. European Commission—Joint Research Centre—Institute for the Protection and Security of the Citizen—Hazard Assessment Unit

CINEAS (2010) Indagine quantitativa su: il livello di consapevolezza rispetto al rischio [...] ed il livello di propensione alla sottoscrizione di un'assicurazione contro i rischi da calamità naturali, Milano, 12 Novembre 2010

de Marchi B, Pellizzoni L, Ungaro D (2001) Il rischio Ambientale, Il Mulino Universale Paperbacks

Deck O, Verdel T (2012) Uncertainties and risk analysis related to geohazards: from practical applications to research trends. In: Emblemsvåg J (ed) Risk management for the future—theory and cases. Intech—www.intechopen.com

Department for International Development—UK aid from the British people (2012) Multi-Hazard disaster risk assessment (v2). Crown, New York

Dall'O' G, Galante A, Ruggieri G (2008) Guida alla valorizzazione energetica degli immobili, Il Sole24Ore

Dall'O G, Norese MF, Galante A, Novello C (2013) A multi-criteria methodology to support public administration decision making concerning sustainable energy action plans. Energies 6:4308–4330

Dovers SR, Handmer JW (1992) Uncertainty, sustainability and change. Glob Environ Change 2(4):262–276

European Commission (EC) (2013) Green paper on the insurance of natural and man-made disasters. Strasbourg. COM 213 final, 16 April 2013

EC—European Commission DG Energy (2013) Energy performance certificates in buildings and their impact on transaction prices and rents in selected EU countries—final report, 19 April 2013

Grünthal G, Thieken AH, Schwarz J, Radtke KS, Smolka A, Merz B (2006) Comparative risk assessments for the city of cologne—storms, floods, earthquakes. Nat Hazards 38:21–44

Erbani F (2010) Il disastro—L'Aquila dopo il terremoto: le scelte e le colpe, Laterza

Firouzabadi S (2007/08) Progetto casa sicura: i processi di certificazione per la sicurezza degli edifici pubblici e privati, Tesi di laurea, Fac di Arch e Soc (Politecnico di Milano), A.A

Fischhoff B, Slovic P, Lichtenstein S (1983) 'The public' vs 'the experts': perceived vs actual disagreements about risks of nuclear power. In: Covello et al (eds) The analysis of actual versus perceived risks. Plenum Press, New York

FOEN-WSL—Swiss Federal Institute for Snow and Avalanche Research SLF (2007) Defense structures in avalanches starting zones. Bern

Foglio A (2008) Il marketing ecologico, FrancoAngeli

Fujita TT (1971) Proposed characterization of tornadoes and hurricanes by area and intensity. SMRP Research Paper, no. 91

Garvan T (2001) Analytical paradigms: the epistemological distances between scientists, policy makers, and the public. Risk Anal 21(3):443–455

Guha-Sapir D, Vos F, Below R, Ponserre S (2011) Annual disaster statistical review 2010: the numbers and trends. Centre for Research on the Epidemiology of Disasters (CRED), Université Catholique de Louvain—Brussels, (disponibile all'indirizzo internet http://www.cred.be)

Geneva Association (2009) The Geneva reports—risk and insurance—the insurance industry and climate change—contribution to the global debate. Geneva (ISSN 1662-3738)

Giannini MS (1979) Rapporto sui principali problemi della Amministrazione dello Stato

Granger K (2003) Quantifying storm tide risk in cairns. Nat Hazards 30:165–185

Gruber Urs (2013) Using GIS for avalanche hazard mapping in Switzerland. WSL Swiss Federal Institute for Snow and Avalanche Research SLF

Grünthal G, Thieken AH, Schwarz J, Radtke KS, Smolka A, Merz B (2006) Comparative risk assessments for the city of cologne—storms, floods, earthquakes. Nat Hazards 38:21–44

Guha-Sapir D, Vos F, Below R, Ponserre S (2011) Annual disaster statistical review 2010: the numbers and trends. Centre for Research on the Epidemiology of Disasters (CRED), Université Catholique de Louvain—Brussels (internet http://www.cred.be)

Hammond A, Adriaanse A, Rodenburg E, Bryant D, Woodward R (1995) Environmental indicators: a systematic approach to measuring and reporting on environmental policy performance in the context of sustainable development. WRI, Washington

Holling CS (1973) Resilience and stability of ecological systems. Annu Rev Ecol Syst 4:1–23

ICSU—International Council for Science (2008) A science plan for integrated research on disaster risk—addressing the challenge of natural and human-induced environmental hazards. ISBN 978-0-930357-66-5

Ishikawa K (Lu D J trans) (1985) What is total quality control? Prentice-Hall Inc, Englewood Cliffs

ISO/IEC 17021 (2011) Conformity Assessment—Requirements for bodies providing audit and certification of management systems, 2nd edn

ISO Guide 73 (2009) Risk management, vocabulary, 1st edn

ISO 19011 (2011) Guidelines for auditing management systems, 2nd edn

ISO 31000 (2009) Risk management—principles and guidelines, 1st edn

Istat—Centro diffusione dati, Conti economici nazionali—anni 2007–2009, 1 marzo 2010, (www.istat.it)

Istat—Direzione centrale comunicazione ed editoria, PIL e indebitamento Amministrazioni pubbliche—anni 2008–2010, 1 marzo 2011, (www.istat.it)

Jacobs K, Williams S (2009) Natural disaster preparation and response: issues for state housing authorities. Australian Housing and Urban Research Institute—Southern Research Centre. Final Report No. 131

Klein RJT, Nicholls RJ, Thomalla F (2003) Resilience to natural hazards: how useful is this concept? Environ Hazards 5:35–45

Kron W (2005) Flood risk = hazard · values · vulnerability. Int Water Resour Assoc—Water Int 30(1):58–68

Kuik O et al (2008) Post-2012 climate policy dilemmas: a review of proposals. Clim Policy 8:317–336

Lenoble C, Durand C (2011) Introduction of frequency in France following the AZF accident. J Loss Prev Process Ind 24 (Elsevier):227–236

Lomberg B (2001) The skeptical environmentalist: measuring the real state of the world. Cambridge University Press, Cambridge

Maccaferri S, Cariboni F, Campolongo F (2012) Natural catastrophes: risk relevance and insurance coverage in the EU. European Commission, Joint Research Centre, Scientific Support to Financial Analysis Unit, Institute for the Protection and Security of the Citizens, Version Sept 2012

Meadows D et al (1992) Beyond the limits: confronting global collapse, envisioning a sustainable future. Chelsea Green, Post Mills

Merz B, Kreibich H, Schwarze R, Thieken A (2010) Assessment of economic flood damage (review article). Nat Hazards Earth Syst Sci 10:1697–1724

Masao Kuroda (1966) Classification of snow avalanches. International conference on low temperature science, I. Conference on physics of snow and ice, II. Conference on cryobiology, August 14–19, 1966. Sapporo, pp 1277–1290

Messner F, Penning-Rowsell E, Green C, Meyer V, Tunstall S, van der Veen A (2007) Evaluating flood damages: guidance and recommendations on principles and methods. FLOODsite Project Deliverable D9.1, contract no: GOCE-CT-2004-505420. www.floodsite.net. Accessed Jan 2007

Montaldo V, Faccioli E, Zonno G, Akinci A, Malagnini L (2005) Treatment of ground-motion predictive relationships for the reference seismic hazard map of Italy. J Seismolog (Springer) 9:295–316

Mumford L (1961) The city in history. Brace and Jovanovich Inc, Harcourt

Munich Re NatCatSERVICE (2013) Natural catastrophes in 2012, Münchener Rückversicherungs-Gesellschaft, Geo Risks Research. (www.munichre.com)

Murphy J, Gouldson A (2000) Environmental policy and industrial innovation: integrating environment and economy through ecological modernization. Geoforum 31:33–44

Neumayer E (2010) Weak versus Strong Sustainability—exploring the limits of two opposing paradigms, 3rd edn. Edward Elgar Publishing, Cheltenham

O'Brien G, O'Keefe P, Rose J, Wisner B (2006) Climate change and disaster management. Disasters 30(1):64–80

OECD (2001) Citizens as partners: information, consultation and public participation in policy-making—a manual

Office of the Deputy Prime Minister (2006) HHSRS Operating Guidance—Housing Act 2004: guidance about inspections and assessment of hazards given under Section 9, London

Oloruntoba R (2013) Plans never go according to plan: an empirical analysis of challenges to plans during the 2009 Victoria bushfires. Technol Forecast Soc Change 80(9):1674–1702. http://dx.doi.org/10.1016/j.techfore.2012.12.002

Pandeya MD, Nathwani JS (2004) Life quality index for the estimation of societal willingness-to-pay for safety. Struct Saf 26:181–199

Penning-Rowsell E, Johnson C, Tunstall S (2006) 'Signals' from pre-crisis discourse: lessons from UK flooding for global environmental policy change? Glob Environ Change 16:323–339

Pérez-Lombard L, Ortiz J, Gonzàlez R, Maestre IL (2009) A review of benchmarking, rating and labelling concepts within the framework of building energy certification schemes (Review). Energ Buildings 41 (Elsevier):272–278

Petrini V (2009) Vulnerability assessment: mutual influence on the procedures for vulnerability and hazard assessment. In: Treu M.C, 2009, Città, Salute, Sicurezza—Strumenti di governo e casi studio—La gestione del rischio, Maggioli Editore, Milano

Pimm SL (1986) Community stability and structure. In: Soule ME (ed) Conservation biology: the science of scarcity and diversity. Sinauer Ass Sunderland, pp 309–329

PRIM, 2007-2010, Programma Regionale Integrato di Mitigazione dei rischi, Regione Lombardia, 2008

Proverbs D, Lamond J (2008) The barriers to resilient rein-statement of flood damaged homes. In: 4th international i-Rec conference, Christchurch, 30th April–2nd May 2008

Ritchey T (2006) Modeling multi-hazard disaster reduction strategies with computer-aided morphological analysis. In: van de Walle B, Turoff M (eds) Proceedings of the 3rd international ISCRAM conference, Newark, May 2006

Rosso R (2005) Criteri per la redazione dei Piani di bacino: ambiti normativi delle fasce di inondabilità in funzione di tiranti idrici e velocità di scorrimento, Regione Liguria

Roy B (1985) Méthodologie multicritère d'aide à la décision, Paris, Economica

Roy B, Bouyssou D (1993) Aide multicritère à la décision : Méthodes et cas, Paris, Economica

Sabetta F (2010) *Valutazione della pericolosità e mitigazione del rischio sismico*, presentazione, Napoli, 5 maggio 2010

Saggin R, Tasca A (2009/10) *Casa sicura 2010*—Estensione della procedura ai rischi idrogeologico e valanghivo: applicazione alla Provincia di Lecco, Elaborato di laurea, Fac di Ing Civile Ambientale e Territoriale (Politecnico di Milano), A.A

Salvati P, Bianchi C, Rossi M, Guzzetti F (2010) Societal landslide and flood risk in Italy. Nat Hazards Earth Syst Sci 10:465–483

Salvi O, Gaston D (2013) Risk assessment in decision making related to land-use planning (LUP) as required by the Seveso II directive, INERIS—"Risk analysis and major accident prevention" project. Verneuil-en-Halatte, Picardy

Siegrist M, Cvetkovich G (2000) Perception of hazards: the role of social trust and knowledge. Risk Anal 20(5):713–720

Simon JL (1981) The ultimate resource. Princeton University Press, Princeton

Schmidt J, Matcham I, Reese S, King A, Bell R, Henderson R, Smart G, Cousins J, Smith W, Heron D (2011) Quantitative multi-risk analysis for natural hazards: a framework for multi-risk modelling. Nat Hazards 58:1169–1192

Sheard S, Mostashari A (2013) A framework for system resilience discussions. Stevens Institute of Technology, Hoboken

Solomos G, Pinto A, Dimova S (2008) A review of the seismic hazard zonation in national building codes in the context of eurocode 8—support to the implementation, harmonization and further development of the Eurocodes, JRC—European Commission

Therivel R (2010) Strategic environmental assessment in action. Earthscan, Oxford

Tietenberg T (2004) Economica dell'ambiente, (trad 2006) a cura di Chiuri MC, D'Amato A, McGraw-Hill. ISBN 88-386-6235-5, Milano

Timmerman P (1981) Vulnerability, resilience and the collapse of society: a review of models and possible climatic applications. Institute for Environmental Studies, University of Toronto, Toronto

Treu MC (2009) Città, salute, sicurezza, Maggioli, Sant'Arcangelo di Romagna, RM

Trigila A, Iadanza C, Vittori E (2007) The WebGIS application of the IFFI Project (Italian landslide inventory). Geophys Res Abs 9:09966

Tronconi O, Ciaramella A, Pisani B (2007) La gestione di edifici e di patrimoni immobiliari, Il Sole 24 Ore, Milano

UNESCO (2009) The international classification for seasonal snow on the ground, prepared by the ICSI-UCCS-IACS. Working group on snow classification IHP-VII Technical Documents in Hydrology No. 83 IACS Contribution No. 1

Unisdr (2007) Hyogo framework for action 2005–2015: building the resilience of nations and communities to disasters. United Nations, Geneva

UN-ISDR (2008) Towards national resilience, good practices of national platforms for disaster risk reduction. United Nations, Geneva

United Nations (2009) Risk and poverty in a changing climate—invest today for a safer tomorrow. global assessment Report on Disaster Risk Reduction (www.preventionweb.net/gar09),

Vertinsky I, Zhou D (2000) Product and process certification—systems, regulations and international marketing strategies. Int Mark Rev 17(3):231–252

Vos F, Rodriguez J, Below R, Guha-Sapir D (2010) Annual disaster statistical review 2009: the numbers and trends. Centre for Research on the Epidemiology of Disasters (CRED), Université Catholique de Louvain – Brussels. (disponibile all'indirizzo internet http://www.cred.be)

Winward J, Schiellerup P, Boardman B (1998) Cool labels: the first three years of the European energy label, environmental change institute. University of Oxford, Oxford

World Bank-International Bank for Reconstruction and Development and United Nations (2010) Natural hazards, unnatural disasters—the economics of effective prevention. ISBN 978-0-8213-8050-5, Washington, D.C.

MIX
Papier aus verantwortungsvollen Quellen
Paper from responsible sources
FSC® C105338

If you have any concerns about our products,
you can contact us on
ProductSafety@springernature.com

In case Publisher is established outside the EU,
the EU authorized representative is:
Springer Nature Customer Service Center GmbH
Europaplatz 3, 69115 Heidelberg, Germany

Printed by Libri Plureos GmbH
in Hamburg, Germany